21st Century Homestead
Organic Fertilizer

Contents

Chapter 1

Organic fertilizer

A cement reservoir containing cow manure mixed with water. This is common in rural Hainan Province, China. Note the bucket on a stick that the farmer uses to apply the mixture.

Organic fertilizers are fertilizers derived from animal matter, human excreta or vegetable matter. (e.g. compost, manure).[1] In contrast, the majority of fertilizers are extracted from minerals (e.g., phosphate rock) or produced industrially (e.g., ammonia). Naturally occurring organic fertilizers include animal wastes from meat processing, peat, manure, slurry, and guano.

Compost bin for small-scale production of organic fertilizer

1.1   Examples and sources of organic fertilizer

The main organic fertilizers are, in ranked order, peat, animal wastes (often from slaughter houses), plant wastes from agriculture, and sewage sludge.[1]

1.1.1   Mineral

The main source of organic fertilizer is peat, an immature precursor to coal. Peat itself offers no nutritional value to the plants, but improves the soil by aeration and absorbing water.

Mined powdered limestone,[2] rock phosphate, and Chilean saltpeter are inorganic (not of biologic origins) compounds, which can be energetically intensive to harvest.[2][3][4]

1.1.2   Animal sources

These materials include the products of the slaughter of animals. Bloodmeal, bone meal, hides, hoofs, and horns are typical precursors.[1]

Chicken litter, which consists of chicken manure mixed with sawdust, is an organic fertilizer that has been shown to better condition soil for harvest than synthesized fertilizer. Researchers at the Agricultural Research Service (ARS) studied the effects of using chicken litter, an organic fertilizer, versus synthetic fertilizers on cotton fields, and found that fields fertilized with chicken litter had a 12% increase in cotton yields over fields fertilized with synthetic fertilizer. In addition

A large commercial compost operation

to higher yields, researchers valued commercially sold chicken litter at a $17/ton premium (to a total valuation of $78/ton) over the traditional valuations of $61/ton due to value added as a soil conditioner.[5]

1.1.3 Plant

Processed organic fertilizers include compost, humic acid, amino acids, and seaweed extracts. Other examples are natural enzyme-digested proteins, fish meal, and feather meal. Decomposing crop residue (green manure) from prior years is another source of fertility.

Other ARS studies have found that algae used to capture nitrogen and phosphorus runoff from agricultural fields can not only prevent water contamination of these nutrients, but also can be used as an organic fertilizer. ARS scientists originally developed the "algal turf scrubber" to reduce nutrient runoff and increase quality of water flowing into streams, rivers, and lakes. They found that this nutrient-rich algae, once dried, can be applied to cucumber and corn seedlings and result in growth comparable to that seen using synthetic fertilizers.[6]

1.1.4 Sewage sludge

Although night soil is a traditional organic fertilizer, the main source of this type is sewage sludge.

Recycled sewage sludge (aka biosolids) as soil amendment is only available to less than 1% of US agricultural land. Industrial pollutants in sewage sludge prevents recycling it as fertilizer. The USDA prohibits use of sewage sludge in organic agricultural operations in the U.S. due to industrial pollution, pharmaceuticals, hormones, heavy metals, and

Peat is the most widely used organic fertilizer.

other factors.[7][8][9] The USDA now requires 3rd-party certification of high-nitrogen liquid organic fertilizers sold in the U.S.[10]

Sewage sludge use in organic agricultural operations in the U.S. has been extremely limited and rare due to USDA prohibition of the practice (due to toxic metal accumulation, among other factors).[11][12][13]

Animal sourced urea and urea-formaldehyde from urine are suitable for organic agriculture; however, synthetically produced urea is not.[14] The common thread that can be seen through these examples is that *organic* agriculture attempts to define itself through minimal processing (e.g., via chemical energy such as petroleum — see Haber process), as well as being naturally occurring or via natural biological processes such as composting.

1.1.5 Examples

- Alfalfa

- Ash[15]

- Blood meal

- Bone meal[16]

- Compost

- Cover crops

Decomposing animal manure, an organic fertilizer source

- Fish emulsion[17]

- Fish meal

- Manure

- Rock phosphate

- Raw Langbeinite

- Rockdust

- Unprocessed natural potassium sulfate

- Wood chips/sawdust[18]

1.2 References

[1] Heinrich Dittmar, Manfred Drach, Ralf Vosskamp, Martin E. Trenkel, Reinhold Gutser, Günter Steffens "Fertilizers, 2. Types" in Ullmann's Encyclopedia of Industrial Chemistry, 2009, Wiley-VCH, Weinheim. doi:10.1002/14356007.n10_n01

[2]

[3] "Can I Use This Input on My Organic Farm?". eXtension. Retrieved 25 August 2010.

[4] Alternative Farming Systems Information Center. "Organic Production and Organic Food: Information Access Tools". Nal.usda.gov. Retrieved 25 August 2010.

[5] "Researchers Study Value of Chicken Litter in Cotton Production". 23 July 2010.

[6] "Algae: A Mean, Green Cleaning Machine". USDA Agricultural Research Service. 7 May 2010.

[7] "Organic Farming | Agriculture | US EPA". Epa.gov. Retrieved 25 August 2010.

[8] "CalOrganic Farms News". Calorganicfarms.com. Retrieved 25 August 2010.

[9] "Biosolids: Targeted National Sewage Sludge Survey Report". EPA.gov. January 2009.

[10] Schrack, Don (23 February 2009). "USDA Toughens Oversight of Organic Fertilizer: Organic fertilizers must undergo testing". The Packer. Retrieved 19 November 2009.

[11] "Organic Farming | Agriculture | US EPA". Epa.gov. Retrieved 2012-01-09.

[12] http://www.ewg.org/reports/sludgememo

[13] "CalOrganic Farms News". Calorganicfarms.com. Retrieved 2012-01-09.

[14] "In a natural organic system, nitrate in the soil is derived from the gradual breakdown of humus". Ecochem.com. Retrieved 2012-01-09.

[15] "Managing Potassium for Organic Crop Production" (PDF). CO State Extension.

[16] "Phosphorus Fertilizers for Organic Farming Systems". CO State Extension.

[17] "Maintaining Soil Fertility in an Organic Fruit and Vegetable Crops System". University of MN Extension.

[18] "Organic Materials as *Nitrogen Fertilizers". CO State Extension.

Chapter 2

Azomite

Azomite mineral ore in its natural state prior to being ground for use

Azomite (pronounced ā-zō-mite, officially all-caps AZOMITE) is the registered trademark for a complex silica ore (hydrated sodium calcium aluminosilicate) with an elevated ratio of trace minerals unique to the Utah mineral deposit from which it is mined. When the ash from a volcanic eruption filled a nearby seabed an estimated 30 million years ago, the combination of seawater, fed by hundreds of mineral-rich rivers and the rare earth minerals present in the volcanic ash, created the deposit's distinctive composition. Mineralogically described as rhyolitic tuff breccia, the geologic characteristic of its surface is referred to as an outcrop known as a hogback.[1]

2.1 Scientific analysis

Scientific analysis of this combination of volcanic ash and marine minerals reports over 70 trace minerals, many recognized as essential by the National Research Council of Canada.[2] A typical analysis of an Azomite sample using spark source mass spectrometry reveals the presence of many rare earth elements.[3]

2.2 History

Rollin J. Anderson, a geological prospector and organic pioneer, founded Azomite in Salt Lake City, Utah in 1942. Convinced that what ailed America was its food supply and the depleted soil from which it came, Anderson left the city life of San Francisco in search of a remedy. He initially pursued development of his father's Utah-based gypsum mine as a means for neutralizing alkaline farmland; however production logistics of a promising start were foiled by World War II crisis. Intrigued by the Native American folklore surrounding the healing powers of the "painted rocks" just south of Salt Lake City, Anderson set out to learn more and to try to validate their claims.

Anderson brought samples of the pink ore to his friend Charles Head, a scientist and chief microscopist at the U.S. Bureau of Mines, whose analysis showed a wide array of minerals similar to the caliche rocks of Chile and Peru, the source for much of the world's nitrate. Head further noted that this was an aluminosilicate mixed with an abundance of minerals, rare in the United States, and even in the world, that appeared to contain "all the essential minerals and trace elements in a balanced ratio and naturally chelated".[4]

The premise of his findings was a "unique" analysis showing over 70 trace minerals, which inspired Anderson to coin the name Azomite for his discovery – an acronym for the A to Z of minerals including trace elements. Selected by the U.S. Government to study nitrates in South America (1919-1925), Head had developed a theory that the benefits plants received from nitrates was actually from the minute quantities of trace elements which served as catalysts. The duo ground the ore into rock dust to see its effects in a controlled environment. Beginning with tomatoes and extending tests to a wide range of vegetables, they found the Azomite plots produced heartier plants that were also free from the worm infestation prevalent in the non-treated plots.[4]

Expanding their studies to livestock and poultry feed provided results showing improvement in animal growth, reproductive vigor and immunity. Poultry studies showed greater egg production with less breakage. Livestock feed costs lessened with Azomite used as a feed additive, enabling less grain expense. In addition, animals showed a definite preference for pasture grown with Azomite and hay from Azomite treated soils. Word spread and based on farmers' testimonials, Anderson mined, crushed and sold Azomite locally, gaining an enthusiastic following of local farmers.[4]

Upon his retirement in 1988 Anderson leased the reserves to mineral mining company Peak Minerals, led by Wes Emerson. Mr. Emerson continued to complete additional accredited scientific studies to support the efficacy of Azomite. In 2011, the Anderson Company's ownership merged with what is now known as Azomite Mineral Products, Inc. under the direction of Mr. Emerson as Company President. Azomite's distribution channels now serve approximately thirty countries, with over thirty percent of the company's sales from international export.[1]

2.3 Use

Azomite is used primarily as a natural feed anticaking agent and remineralizer for depleted soils. Hydrated Sodium Calcium Aluminosilicate (HSCAS), its primary component, is listed in the U.S. Code of Federal Regulations (21 CFR 582.2729) as an anticaking agent for livestock feed and is generally recognized as safe (GRAS) by the U.S. Food and Drug Administration (FDA). Agriculture and livestock producers have used Azomite to support livestock health and plant nutrition for over seventy years. While it contains minute quantities of naturally occurring contaminants, Azomite falls well within the guidelines for use in animal feed by the Association of American Feed Control Officials.

As a natural substance, Azomite is listed by the Organic Materials Review Institute (OMRI) for use in organic farming.[5] Adequately mineralized soil has a natural resilience to pests, supporting a reduced dependency on pesticides and fertilizers. In addition, evolving studies link organic crops with higher nutritional levels than those produced by non-organic means.[6]

In 1997, Jared Milarch, a horticulture student at Northwestern Michigan College introduced Azomite to the nursery industry after testing the rock dust on plant growth in controlled studies on his family's chemical-free commercial farm. Milarch conducted experiments proving his theory that Azomite works as a catalyst to help plants better absorb nutrients from the soil.[7]

Independent scientist Lee Klinger believes soil acidification is among the leading factors contributing to sudden oak death, by altering mineral balance and reducing availability of nutrients. Based on this theory, he has developed a holistic treatment to successfully treat the disease that includes a regimen of Azomite.[8]

Azomite is not approved by the FDA for human consumption. Weston A. Price Foundation founder, Sally Fallon, considers Azomite's bioavailable trace mineral content to be a superfood; beneficial to human health.[9]

2.4 References

[1] Redgrave, Chris (3 May 2012). "Zions Bank Speaking on Business: Azomite Mineral Products" (AUDIO). *KSL Radio, Utah*. Retrieved 9 September 2012.

[2] "Minerals for Plants, Animals and Man". *Alberta Agriculture and Rural Development*. Government of Alberta. Retrieved 5 October 2012.

[3] "Certificate of Analysis" (PDF). *Azomite Company Website*. Retrieved 1 October 2012.

[4] Tompkins, Peter; Bird, Christopher (2002). "Chapter 17, Savory Soil". *Secrets of the Soil* (Third ed.). Earthpulse Press (originally published by Harper & Row). ISBN 1-890693-24-3.

[5] "Azomite Mineral Products Inc.". *Organic Materials Review Institute*. OMRI. Retrieved 25 September 2012.

[6] Palmer, RD, Sharon (July 2009). "Digging Into Soil Health". *Today's Dietitican* 11 (7): 38. Retrieved 5 October 2012.

[7] Yarrow, David (2000). "Mineral Restoration & Utah Rock Dust" (PDF). *ACRES Magazine, A Voice for Eco-Agriculture* 30 (4): 14–17. Retrieved 5 October 2012.

[8] Rich, Deborah K. (October 8, 2005). "OAK LORE / Preserving a heritage tree / Scientist takes holistic approach to sudden oak death". *San Francisco Chronicle*. Retrieved 5 October 2012.

[9] Fallon, Sally; Enig,Ph.D., Mary G. (2001). *Nourishing Traditions* (Second ed.). New Trends Publishing, Inc. p. 617. ISBN 978-09670897-3-7.

2.5 External links

- U.S. Company website
- International Company website

Chapter 3

Bioeffector

A **Bioeffectors** is a viable microorganism or active natural compounds which directly or indirectly affects plant performance (Biofertilizer), and thus has the potential to reduce fertilizer and pesticide use in crop production.[1]

3.1 Types

Bioeffectors have a direct or indirect effect on plant performance by influencing the functional implementation or activation of biological mechanisms, particularly those interfering with soil-plant-microbe interactions.[2] In contrast to conventional fertilizers and pesticides, the effectiveness of bioeffectors is not based on a substantial direct input of mineral plant nutrients, either in inorganic or organic forms.

- Products in use are:

    - Microbial residues,

    - Composting and fermentation products,

    - Plant and algae extracts

- Bioeffector-preparations (*bio-agents*) as ready-formulated products are applied:

    - with the purpose of stimulating plant growth (bio-stimulants),

    - to improve plant nutrient acquisition (bio-fertilizers),

    - to protect plants from pathogens and pests (bio-control agents)

    - or generally to advance cropping efficiency; they can contain one or more bio-effectors along with other substances"[3]

- Well established bioeffectors with documented positive results in the field level are:

    - Rhizobia strains for soil or seed inoculation as a prerequisite for symbiotic N2-fixation when establishing new legume species or varieties.

    - positive effects of mycorrhiza inoculation for soils with a (temporarily) low potential for natural root mycorrhization.

    - sufficient mycorrhization enhances nutrient (P) and water uptake and increases resistance to pathogenic fungi.

- Further mechanisms for the positive impact of bioeffectors on plant growth have postulated, promising a high potential for resource preservation due to reduction of fertiliser and pesticide use:

- Active nutrient mobilisation by exudation of acids and carboxylates (e.g. P-mobilisation),

- exudation of micro-nutrient mobilising siderophores/chelates (e.g. Fe3+),

- reduction of trace elements from less soluble oxidised to highly soluble reduced forms (e.g. Fe3+ to Fe2+, Mn4+ to Mn2+),

- associative/non-symbiotic N2-fixation, protective antagonism to plant pathogens,

- enhancement of mycorrhizal infection and growth, and stimulating hormonal effects.

3.2 Research and Public Dissemination

Under the Acronym *Biofector* the European Union supports the Research of Bioeffectors under the leadership of the University of Hohenheim (Coordinator Guenter Neumann).[4] The results of the project will be evaluated by the members of the Association Biostimulants in Agriculture (ABISTA) and provided agriculture for use and EU institutions for the legislative and registration procedures.[5]

3.3 External links

- Webpage Biofector

- Webpage Association Biostimulants in Agriculture

- Webpage Biofector CULS Prague

- Webpage Madora Bioeffectors

- Webpage Biofector University of Hohenheim

3.4 References

[1] Minutes of the 6th International Symposium Plant Protection and Plant Health in Europe, May 2014 Braunschweig, Germany

[2] V. Römheld, G. Neumann (2006): *The Rhizosphere: Contributions of the soil-root interface to sustainable soil systems*. In: N. Uphoff, N., N. A. S. Ball et al. (Hg.), *Biological Approaches to Sustainable Soil Systems*, S. 92–107, CRC-Press, Oxford, UK.

[3] Bakonyi N., Donath S., Weinmann M., Neumann G., Müller T., Römheld V. (2008): Assessing commercial bio-fertilisers for improved phoshorus availability. Use of rapid screening tests. Jahrestagung der Deutschen Gesellschaft für Pflanzenernährung 2008

[4] European Research Program Biofector

[5] Webpage Biostimulants Association

Chapter 4

Biofertilizer

A **Bio fertilizer** (also *bio-fertilizer*) is a substance which contains living microorganisms which, when applied to seed, plant surfaces, or soil, colonizes the rhizosphere or the interior of the plant and promotes growth by increasing the supply or availability of primary nutrients to the host plant.[1] Bio-fertilizers add nutrients through the natural processes of nitrogen fixation, solubilizing phosphorus, and stimulating plant growth through the synthesis of growth-promoting substances. Bio-fertilizers can be expected to reduce the use of chemical fertilizers and pesticides. The microorganisms in bio-fertilizers restore the soil's natural nutrient cycle and build soil organic matter. Through the use of bio-fertilizers, healthy plants can be grown, while enhancing the sustainability and the health of the soil. Since they play several roles, a preferred scientific term for such beneficial bacteria is "plant-growth promoting rhizobacteria" (PGPR). Therefore, they are extremely advantageous in enriching soil fertility and fulfilling plant nutrient requirements by supplying the organic nutrients through microorganism and their byproducts. Hence, bio-fertilizers do not contain any chemicals which are harmful to the living soil.

Bio-fertilizers provide eco-friendly organic agro-input and are more cost-effective than chemical fertilizers. Bio-fertilizers such as Rhizobium, Azotobacter, Azospirilium and blue green algae (BGA) have been in use a long time. Rhizobiuminoculant is used for leguminous crops. Azotobacter can be used with crops like wheat, maize, mustard, cotton, potato and other vegetable crops. Azospirillum inoculations are recommended mainly for sorghum, millets, maize, sugarcane and wheat. Blue green algae belonging to a general cyanobacteria genus, *Nostoc* or *Anabaena* or *Tolypothrix* or *Aulosira*, fix atmospheric nitrogen and are used as inoculations for paddy crop grown both under upland and low-land conditions. *Anabaena* in association with water fern Azolla contributes nitrogen up to 60 kg/ha/season and also enriches soils with organic matter.[2]

Other types of bacteria, so-called phosphate-solubilizing bacteria, such as Pantoea agglomerans strain P5 or Pseudomonas putida strain P13,[3] are able to solubilize the insoluble phosphate from organic and inorganic phosphate sources.[4] In fact, due to immobilization of phosphate by mineral ions such as Fe, Al and Ca or organic acids, the rate of available phosphate (P_i) in soil is well below plant needs. In addition, chemical P_i fertilizers are also immobilized in the soil, immediately, so that less than 20 percent of added fertilizer is absorbed by plants. Therefore, reduction in P_i resources, on one hand, and environmental pollutions resulting from both production and applications of chemical P_i fertilizer, on the other hand, have already demanded the use of new generation of phosphate fertilizers globally known as phosphate-solubilizing bacteria or phosphate bio-fertilizers.

4.1 Benefits

A bio-fertilizer provides the following benefits:

1. Since a bio-fertilizer is technically living, it can symbiotically associate with plant roots. Involved microorganisms could readily and safely convert complex organic material in simple compounds, so that plants are easily taken up. Microorganism function is in long duration, causing improvement of the soil fertility. It maintains the natural

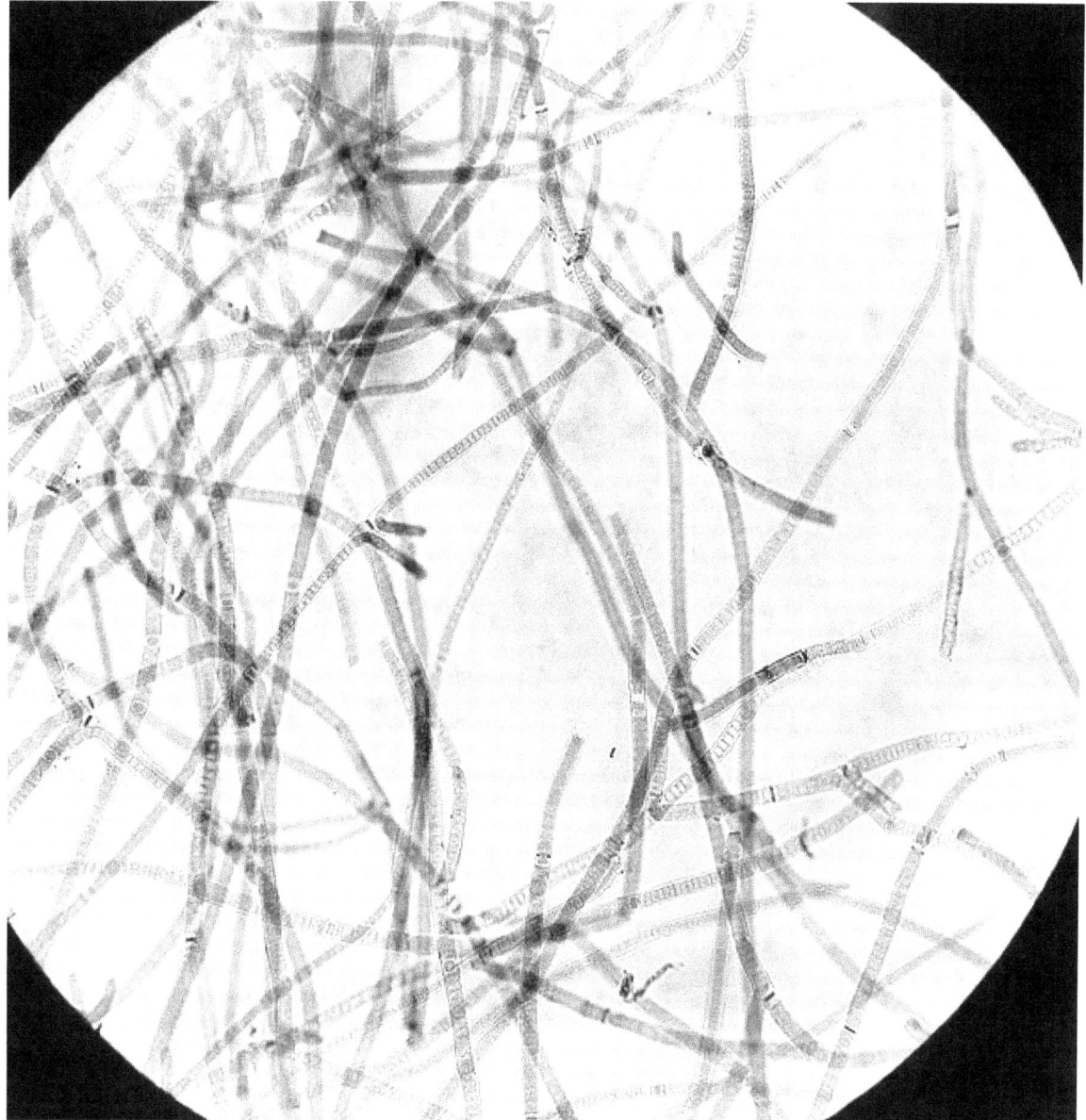

Tolypothrix*, cyanobacteria often used as fertilizer*

habitat of the soil. It increases crop yield by 20-30%, replaces chemical nitrogen and phosphorus by 25%, and stimulates plant growth. It can also provide protection against drought and some soil-borne diseases.

2. Bio-fertilizers are cost-effective relative to chemical fertilizers. They have lower manufacturing costs, especially regarding nitrogen and phosphorus use.

Some important groups of Bio-fertilizers

1. Azolla-Anabena symbiosis: Azolla is a small, eukaryotic, aquatic fern having global distribution.Prokaryotic blue green algae Anabena azolla resides in its leaves as a symbiont. Azolla is an alternative nitrogen source. This association has gained wide interest because of its potential use as an alternative to chemical fertilizers.

2. Rhizobium: Symbiotic nitrogen fixation by Rhizobium with legumes contribute substantially to total nitrogen fixation. Rhizobium inoculation is a well-known agronomic practice to ensure adequate nitrogen.

Blue-green algae cultured in specific media. Blue-green algae can be helpful in agriculture as they have the -green algae is used it as a bio-fertilizer.

4.2 See also

- Endophyte
- Microbial inoculant

4.3 References

[1] Vessey, J.k. 2003, Plant growth promoting rhizobacteria as bio-fertilizers. Plant Soil 255, 571-586

[2] http://eprints.ru.ac.za/36/1/Kiguli.PDF

[3] http://www.springerlink.com/content/v2315pl5736061g7/fulltext.pdf

[4] http://www.springerlink.com/content/q327j346t7233222/fulltext.pdf

4.4 External links

- greenbiotech-co.com
- Foliar BioFertilizer Recipe

Chapter 5

Biosolids

Pumpkin seedlings planted out on windrows of composted biosolids

Biosolids is a term coined in the United States that is typically used to describe several forms of treated sewage sludge that is intended for agricultural use as a soil conditioner. Although sewage sludge has long been used in agriculture [needs citation], concerns about offensive odors and disease risks from pathogens and toxic chemicals may reduce public acceptance of the practice. Modern use of the term *biosolids* may be subject to government regulations, although informal use describes a broader range of semi-solid organic products separated from sewage.

Description of *biosolids* in conformance with local regulations may reduce confusion; but some use an expanded definition including any solids, slime solids or liquid slurry residue generated during the treatment of domestic sewage including scum and solids removed during primary, secondary or advanced treatment processes. Use of alternative terms like *solids* or *wastewater solids* may be preferable for non-conforming biosolids.[1]

5.1   Terminology

Biosolids may be defined as organic wastewater solids that can be reused after suitable sewage sludge treatment processes leading to sludge stabilization such as anaerobic digestion and composting.[2]

Alternatively, the biosolids definition may be restricted by local regulations to wastewater solids only after those solids have completed a specified treatment sequence and/or have concentrations of pathogens and toxic chemicals below specified levels.[3]

The United States Environmental Protection Agency (USEPA) defines the two terms - sewage sludge an biosolids - in the Code of Federal Regulations (CFR), Title 40, Part 503 as follows: *sewage sludge* refers to the solids separated during the treatment of municipal wastewater (including domestic septage), while *biosolids* refers to treated sewage sludge that meets the USEPA pollutant and pathogen requirements for land application and surface disposal.[3] A similar definition has been used internationally.[4]

5.2   Characteristics

5.2.1   Quantities

Approximately 7,100,000 dry tons of biosolids were generated in 2004 at approximately 16,500 municipal wastewater treatment facilities in the United States.[5]

In the United States, as of 2013 about 55% of sewage solids are turned into fertilizer, despite demand from farmers who wish to buy more.[6] Challenges to increased levels of recycling include capital needed to build digesters, the complexity of complying with health regulations, and avoiding neighbors who object to unpleasant smells. There are also new forms of contaminants in urban sewage systems which make the process of producing high quality biosolids more complex. These have led some municipalities to ban biosolids on farms and even in forests.

5.2.2   Nutrients

Encouraging agricultural use of biosolids is intended to prevent filling landfills with nutrient-rich organic materials from the treatment of domestic sewage that might be recycled and applied as fertilizer to improve and maintain productive soils and stimulate plant growth.[5] Biosolids may contain macronutrients nitrogen, phosphorus, potassium and sulphur with micronutrients copper, zinc, calcium, magnesium, iron, boron, molybdenum and manganese.[4]

5.2.3   Industrial and man-made contaminants

The United States Environmental Protection Agency (USEPA) and others have shown that biosolids can contain measurable levels of synthetic organic compounds, radionuclides and heavy metals.[4][7][8] USEPA has set numeric limits for arsenic, cadmium, copper, lead, mercury, molybdenum, nickel, selenium, and zinc but has not regulated dioxin levels.[5][9]

So-called "contaminates of emerging concern" may also be present in biosolids.[10] The United States Geological Survey analyzed in 2014 nine different consumer products containing biosolids as a main ingredient for 87 organic chemicals found in cleaners, personal care products, pharmaceuticals, and other products. These analysis detected 55 of the 87 organic chemicals measured in at least one of the nine biosolid samples, with as many as 45 chemicals found in a single sample.[11]

5.2.4 Pathogens

In the United States the USEPA mandates certain treatment processes designed to significantly decrease levels of certain so-called indicator organisms, in biosolids.[5] These include, "...operational standards for fecal coliforms, *Salmonella* sp. bacteria, enteric viruses, and viable helminth ova."[12]

However, the US-based Water Environment Research Foundation has shown that some pathogens do survive sewage sludge treatment.[13]

USEPA has also classified other pathogens that can appear in biosolids such as various protozoa, bacteria, viruses, and prions as "pathogens of emerging concern".[14] EPA regulations allow only biosolids with no detectable pathogens to be widely applied; those with remaining pathogens are restricted in use.[15]

5.3 Classification systems

5.3.1 United States

In the United States Code of Federal Regulations (CFR), Title 40, Part 503 governs the management of biosolids. Within that federal regulation biosolids are generally classified differently depending upon the quantity of pollutants they contain and the level of treatment they have been subjected to (the latter of which determines both the level of vector attraction reduction and the level of pathogen reduction). These factors also affect how they may be disseminated (bulk or bagged) and the level of monitoring oversight which, in turn determines where and in what quantity they may be applied.[16]

5.4 Alternative terms for similar materials

- Sludge, broadly describes any semi-solid slurry.
- Sewage sludge, broadly describes sludges removed from sewage or municipal wastewater, including:
  - Primary sludge, removed during primary treatment in a primary clarifier
  - Secondary sludge, removed following secondary treatment by biochemical oxidation
  - Digestate, primary and/or secondary sludge following aerobic or anaerobic digestion
  - Thickened sludge, sludge following water removal in a thickener

5.5 History

Further information: Sewage sludge § History

As public concern arose about disposal in the United States of increasing volumes of solids being removed from sewage during sewage treatment mandated by the Clean Water Act, the Water Environment Federation (WEF) sought a new name to distinguish the clean, agriculturally viable product generated by modern wastewater treatment from earlier forms of sewage sludge widely remembered for causing offensive or dangerous conditions. Of three-hundred suggestions, *biosolids* was attributed to Dr. Bruce Logan of the University of Arizona, and recognized by WEF in 1991.[17]

5.6 Examples

- Milorganite is the trademark of a biosolids fertilizer produced by the Milwaukee Metropolitan Sewerage District.[18] The recycled organic nitrogen fertilizer from the Jones Island Water Reclamation Facility in Milwaukee, Wisconsin, is sold throughout North America, reduces the need for manufactured nutrients.

- Loop is the trademark of a biosolids soil amendment produced by the King County Wastewater Treatment Division.[19] Loop has been blended into GroCo, a commercially available compost product, since 1976. Several local farms and forests also use Loop directly.

- TAGRO is short for "Tacoma Grow" and is produced by the City of Tacoma, Washington since 1991.[20][21]

5.7   References

[1] Turovskiy, Izrail S. "Biosolids or Sludge? The Semantics of Terminology". Water and Wastes Digest. Retrieved 24 April 2015.

[2] *Wastewater engineering : treatment and reuse* (4th ed.). Metcalf & Eddy, Inc., McGraw Hill, USA. 2003. p. 1449. ISBN 0-07-112250-8.

[3] "Sewage Sludge/Biosolids Program". United States Environmental Protection Agency. Retrieved 24 April 2015.

[4] "What are biosolids?". Australian Water Association. Retrieved 24 April 2015.

[5] "Questions and Answers on Land Application of Biosolids" (PDF). Water Environment Federation. Retrieved 24 April 2015.

[6] Cities Turn Sewage Into 'Black Gold' For Local Farms (2013)

[7] "Biosolids: Targeted National Sewage Sludge Survey Report - Overview | Biosolids | US EPA". *water.epa.gov*. Retrieved 2015-05-18.

[8] "ISCORS Assessment of Radioactivity in Sewage Sludge: Recommendations on Management of Radioactive Materials in Sewage Sludge and Ash at Publicly Owned Treatment Works" (PDF). *United States Environmental Protection Agency (EPA)*. Interagency Steering Committee on Radiation Standards. April 2004. Retrieved 18 May 2015.

[9] "Final Action Not to Regulate Dioxins in Land-Applied Sewage Sludge | Biosolids | US EPA". *water.epa.gov*. Retrieved 2015-05-18.

[10] "Analytical Methods: Contaminants of Emerging Concern | Pharmaceuticals & Personal Care Products | US EPA". *water.epa.gov*. Retrieved 2015-05-18.

[11] "Land Application of Municipal Biosolids". *Environmental Health - Toxic Substances*. United States Geological Survey. Retrieved 24 April 2015.

[12] *Biosolids Applied to Land: Advancing Standards and Practices*. National Academy of Sciences. 2002. p. 22. ISBN 0-309-08486-5.

[13] "Assessing the Fate of Emerging Pathogens in Biosolids". *Water Environment Research Foundation*. Retrieved 2015-05-18.

[14] "http://water.epa.gov/scitech/research-riskassess/researchstrategy/upload/compendium.pdf" (PDF). *water.epa.gov*. p. xxxv. Retrieved 2015-05-18.

[15] "Biosolids FAQ, Questions 17-18". *water.epa.gov*. Retrieved 2015-06-21.

[16] "A Plain English Guide to the EPA Part 503 Biosolids Rule, Chapter 2 "Land Application of Biosolids"" (PDF). *water.epa.gov*. p. 31. Retrieved 2015-05-20.

[17] "Biosolids: A Short Explanation and Discussion" (PDF). *WEF/U.S. EPA Biosolids Fact Sheet Project*. Water Environment Federation. Retrieved 24 April 2015.

[18] "About us". Milorganite/Milwaukee Metropolitan Sewerage District. Retrieved 27 April 2015.

[19] "What is Loop?". King County Wastewater Treatment Division. Retrieved 20 June 2015.

[20] "About TAGRO". City of Tacoma. Retrieved 20 June 2015.

[21] http://www.cityoftacoma.org/cms/one.aspx?objectId=16884

Testing for human pathogens in cereal crops after the application of biosolids. Biosolids are applied as fertilizer in the Central Wheatbelt of Australia as a recycling program by the Water Corporation.

Chapter 6

Chicken manure

This article is about organic fertilizer. For the slang term, see Chicken shit.

Chicken manure is the feces of chickens used as an organic fertilizer, especially for soil low in nitrogen.[1] Of all animal

A chicken manure sample being collected for a nutrient analysis

manures, it has the highest amount of nitrogen, phosphorus, and potassium.[2] Chicken manure is sometimes pelletized for use as a fertilizer, and this product may have additional phosphorus, potassium or nitrogen added.[3] Optimal storage conditions for chicken manure includes it being kept in a covered area and retaining its liquid, because a significant amount of nitrogen exists in the urine.[4]

Fresh chicken manure contains approximately 1.5% nitrogen.[5] One chicken produces approximately 8-11 pounds of

A poultryman in 1943 on a Hampshire County farm in England moves a poultry fold into line with the others in the field. Each of these chicken sheds contains 25 birds and are moved their length every day, providing fresh ground for the hens to feed on and also ensuring that the chicken manure is spread across the whole field.

manure monthly.[5] Chicken manure can be used to create homemade plant fertilizer.[5]

6.1 Studies

In 1986, a master's thesis study in the Philippines compared the effects of using various fertilizers to enhance milkfish production in brackish water ponds.[6] The study compared the use of using chicken manure only, cow manure only, 16-20-0 fertilizer only, a mixture of cow manure and 16-20-0 fertilizer, a mixture of chicken manure and 16-20-0 fertilizer, and a control group that used no fertilizer.[6] The study concluded that the use of cow manure only as a fertilizer fared

A manure car for the transport of chicken manure at a chicken house in Dolores, Colorado

best, and the use of chicken manure only as a fertilizer fared second best.[6]

6.2 Pollution

Mass applications of chicken manure may create an unpleasant odor. In April 2014 in Escondido, California, a golf course that had "dumped" chicken manure on its grounds was cited by the county government after complaints from local residents about the odor.[7]

In December 2011, the environmental group Environment Maryland asserted that water runoff from agricultural land fertilized with chicken manure was increasing the pollution levels of Chesapeake Bay.[8] The group asserted that excessive phosphorus from the runoff was contributing to the increase of dead zones in the bay.[8] In 2015, in efforts to address the matter before leaving office, Maryland Governor Martin O'Malley put a new regulation into use that "would have limited the amount of poultry manure that Eastern Shore farmers can use on their fields".[9] However, the following Governor Larry Hogan quickly absolved the new regulation after being sworn into office.[9] The runoff problem has been attributed to the use of "an outdated scientific tool for calculating the correct amount of manure".[9] A proposed solution from scientists at the University of Maryland is to have farmers use a new (corrected) formula to calculate proper quantities of chicken manure for agricultural uses.[9]

Chicken sheds at Balado Airfield, Scotland. Poultry sheds like this are common in the Kinross area of Scotland. Manure from the sheds is now collected for use as fuel in a biomass-burning power station at Westfield in Fife.

6.2.1 Human deterrent

Chicken manure has been used as a human deterrent. In July 2013 in Abbotsford, British Columbia, city workers applied chicken manure at a tent encampment to deter homeless people from the area.[10] The affected homeless planned on initiating small claims lawsuits for loss of property and property damage.[10] One of the affected homeless people described the tactics of city workers as "a chicken shit way to do things".[11] The mayor of Abbotsford and the Fraser Valley city manager later apologized regarding the incident.[10][12] Similar instances of using chicken manure in this manner have occurred in British Columbia in Surrey and in Port Coquitlam, the latter of which occurred "shortly after the Abbotsford incident".[10]

6.3 See also

- Chicken shit
- Guano
- Liquid manure
- Manure spreader
- Plant nutrition
- Agriculture and Agronomy portal

6.4   References

[1] Telkamp, Mick. "The Straight Poop On Using Chicken Manure as Fertilizer". Retrieved 16 February 2015.

[2] Deborah L. Martin; Grace Gershuny, eds. (1992). *The Rodale Book of Composting: Easy Methods for Every Gardener* (revised ed.). Rodale. p. 126. ISBN 9780878579914.

[3] Barrett, J. (2008). *FCS Soil Science L3*. FET college series. Pearson Education South Africa. p. 70. ISBN 978-1-77025-114-4.

[4] Pullin, R.S.V.; Shehadeh, Z.H. (1980). *Integrated Agriculture-aquaculture Farming Systems: Proceedings of the ICLARM-SEARCA Conference on Integrated Agriculture-Aquaculture Farming Systems, Manila, Philippines, 6-9 August 1979*. ICLARM conference proceedings. International Center for Living Aquatic Resources Management. p. 80.

[5] Foreman, Patricia; Long, Cheryl (April-May 2013). "Chickens in the Garden: Eggs, Meat, Chicken Manure Fertilizer and More". *Mother Earth News*. Retrieved February 18, 2015.

[6] Garcia, Y.T.; Aragon, C.T.; Dator, M.A.L. *Milkfish Bibliography A Compilation of Abstracts on Milkfish Studies*. WorldFish. p. 191.

[7] "Chicken manure stink could be costly". *U-T San Diego*. April 14, 2014. Retrieved February 18, 2015.

[8] "Chicken manure adds to Chesapeake Bay pollution, group says". *WTOP*. December 28, 2011. Retrieved February 18, 2015.

[9] "Hogan shelves chicken manure rules". *The Frederick News-Post*. January 27, 2015. Retrieved February 18, 2015.

[10] "Big Stink Over Manure Dump". *The Huffington Post*. July 24, 2013. Retrieved February 18, 2015.

[11] "Abbotsford Homeless Campers Clash With City, Police". *The Huffington Post*. June 19, 2013. Retrieved February 18, 2015.

[12] "Chicken Poop Scheme Shames Mayor". *The Huffington Post*. June 6, 2013. Retrieved February 18, 2015.

6.5   Further reading

- Raston, Kate (January 8, 2015). "A new use for chicken manure". *The West Australian*. Retrieved February 18, 2015.

- Patience, Martin (June 25, 2012). "Poultry power: Turning chicken manure to energy". BBC News. Retrieved February 18, 2015.

- "Are humans endangered if cattle dine on chicken manure?". CNN. August 23, 1997. Retrieved February 18, 2015.

- Dabney, Seth Mason (May 1978). *Chicken manure in New York State* (volume 1). Cornell University.

Chapter 7

Compost

A community-level composting plant in a rural area in Germany

Compost (/ˈkɒmpɒst/ or /ˈkɒmpoʊst/) is organic matter that has been decomposed and recycled as a fertilizer and soil amendment. Compost is a key ingredient in organic farming. At the simplest level, the process of composting simply requires making a heap of wetted organic matter known as green waste (leaves, food waste) and waiting for the materials to break down into humus after a period of weeks or months. Modern, methodical composting is a multi-step, closely monitored process with measured inputs of water, air, and carbon- and nitrogen-rich materials. The decomposition process is aided by shredding the plant matter, adding water and ensuring proper aeration by regularly turning the mixture. Worms and fungi further break up the material. Bacteria requiring oxygen to function (aerobic bacteria) and fungi manage the chemical process by converting the inputs into heat, carbon dioxide and ammonium. The ammonium (NH_4) is the form of nitrogen used by plants. When available ammonium is not used by plants it is further converted by bacteria into nitrates (NO_3) through the process of nitrification.

Compost is rich in nutrients. It is used in gardens, landscaping, horticulture, and agriculture. The compost itself is beneficial for the land in many ways, including as a soil conditioner, a fertilizer, addition of vital humus or humic acids, and as a natural pesticide for soil. In ecosystems, compost is useful for erosion control, land and stream reclamation, wetland construction, and as landfill cover (see compost uses). Organic ingredients intended for composting can alternatively be used to generate biogas through anaerobic digestion.

7.1   Terminology

The term "composting" is used worldwide with differing meanings. Some composting textbooks narrowly define composting as being an aerobic form of decompostion, primarily by microbes. For many people, however, composting is used to refer to several different types of biological process. In North America, "anaerobic composting" is still a common term, but in much of the rest of the world and in technical publications the more accurate term anaerobic digestion is used as the microbes used and the processes involved are quite different.

7.2   Ingredients

Home compost barrel in the Escuela Barreales, Santa Cruz, Chile.

7.2.1 Carbon, nitrogen, oxygen, water

Materials in a compost pile.

Composting organisms require four equally important ingredients to work effectively:

- Carbon — for energy; the microbial oxidation of carbon produces the heat, if included at suggested levels.[1]

    - High carbon materials tend to be brown and dry.

- Nitrogen — to grow and reproduce more organisms to oxidize the carbon.

    - High nitrogen materials tend to be green (or colorful, such as fruits and vegetables) and wet.[2]

- Oxygen — for oxidizing the carbon, the decomposition process.

- Water — in the right amounts to maintain activity without causing anaerobic conditions.

Certain ratios of these materials will provide beneficial bacteria with the nutrients to work at a rate that will heat up the pile. In that process much water will be released as vapor ("steam"), and the oxygen will be quickly depleted, explaining the need to actively manage the pile. The hotter the pile gets, the more often added air and water is necessary; the air/water balance is critical to maintaining high temperatures ($135°-160°$ Fahrenheit / $50°$ - $70°$ Celsius) until the materials are broken down. At the same time, too much air or water also slows the process, as does too much carbon (or too little nitrogen).

The most efficient composting occurs with an optimal carbon:nitrogen ratio of about 10:1 to 20:1.[3] Nearly all plant and animal materials have both carbon and nitrogen, but amounts vary widely, with characteristics noted above (dry/wet,

Food scraps compost heap.

brown/green).[4] Fresh grass clippings have an average ratio of about 15:1 and dry autumn leaves about 50:1 depending on species. Mixing equal parts by volume approximates the ideal C:N range. Few individual situations will provide the ideal mix of materials at any point. Observation of amounts, and consideration of different materials[5] as a pile is built over time, can quickly achieve a workable technique for the individual situation.

7.2.2 Animal manure and bedding

On many farms, the basic composting ingredients are animal manure generated on the farm and bedding. Straw and saw-dust are common bedding materials. Non-traditional bedding materials are also used, including newspaper and chopped cardboard. The amount of manure composted on a livestock farm is often determined by cleaning schedules, land avail-ability, and weather conditions. Each type of manure has its own physical, chemical, and biological characteristics. Cattle and horse manures, when mixed with bedding, possess good qualities for composting. Swine manure, which is very wet and usually not mixed with bedding material, must be mixed with straw or similar raw materials. Poultry manure also must be blended with carbonaceous materials - those low in nitrogen preferred, such as sawdust or straw.[6]

7.2.3 Microorganisms

With the proper mixture of water, oxygen, carbon, and nitrogen, micro-organisms are allowed to break down organic matter to produce compost. The composting process is dependent on micro-organisms to break down organic matter into compost. There are many types of microorganisms found in active compost of which the most common are:[7]

- Bacteria- The most numerous of all the microorganisms found in compost. Depending on the phase of composting, mesophilic or thermophilic bacteria may predominate.

- Actinobacteria- Necessary for breaking down paper products such as newspaper, bark, etc.

- Fungi- Molds and yeast help break down materials that bacteria cannot, especially lignin in woody material.

- Protozoa- Help consume bacteria, fungi and micro organic particulates.

- Rotifers- Rotifers help control populations of bacteria and small protozoans.

In addition, earthworms not only ingest partly composted material, but also continually re-create aeration and drainage tunnels as they move through the compost.

A lack of a healthy micro-organism community is the main reason why composting processes are slow in landfills with environmental factors such as lack of oxygen, nutrients or water being the cause of the depleted biological community.[7]

Phases of composting

Under ideal conditions, composting proceeds through three major phases:[7]

- An initial, mesophilic phase, in which the decomposition is carried out under moderate temperatures by mesophilic microorganisms.

- As the temperature rises, a second, thermophilic phase starts, in which in decomposition is carried out by various thermophilic bacteria under high temperatures.

- As the supply of high-energy compounds dwindles, the temperature starts to decrease, and the mesophiles once again predominate in the maturation phase.

7.2.4 Human waste

Human waste (excreta) can also be added as an input to the composting process, like it is done in composting toilets, as human waste is a nitrogen-rich organic material.

People excrete far more water-soluble plant nutrients (nitrogen, phosphorus, potassium) in urine than in feces.[8] Human urine can be used directly as fertilizer or it can be put onto compost. Adding a healthy person's urine to compost usually will increase temperatures and therefore increase its ability to destroy pathogens and unwanted seeds. Urine from a person with no obvious symptoms of infection is much more sanitary than fresh feces. Unlike feces, urine does not attract disease-spreading flies (such as house flies or blow flies), and it does not contain the most hardy of pathogens, such as parasitic worm eggs. Urine usually does not stink for long, particularly when it is fresh, diluted, or put on sorbents.

Urine is primarily composed of water and urea. Although metabolites of urea are nitrogen fertilizers, it is easy to over-fertilize with urine, or to utilize urine containing pharmaceutical (or other) content, creating too much ammonia for plants to absorb, acidic conditions, or other phytotoxicity.[9]

Humanure

"Humanure" is a combination of the words *human* and *manure*, designating human excrement (feces and urine) that is recycled via composting for agricultural or other purposes. The term was first used in a 1994 book by Joseph Jenkins that advocates the use of this organic soil amendment.[10] The term humanure is used by compost enthusiasts in the US but not generally elsewhere. Because the term "humanure" has no authoritative definition it is subject to various uses; news reporters occasionally fail to correctly distinguish between humanure and sewage sludge or "biosolids".[11]

7.3 Uses

Main article: Uses of compost

Compost is generally recommended as an additive to soil, or other matrices such as coir and peat, as a tilth improver, supplying humus and nutrients. It provides a rich *growing medium*, or a porous, absorbent material that holds moisture and soluble minerals, providing the support and nutrients in which plants can flourish, although it is rarely used alone, being primarily mixed with soil, sand, grit, bark chips, vermiculite, perlite, or clay granules to produce loam. Compost can be tilled directly into the soil or growing medium to boost the level of organic matter and the overall fertility of the soil. Compost that is ready to be used as an additive is dark brown or even black with an earthy smell.[12]

Generally, direct seeding into a compost is not recommended due to the speed with which it may dry and the possible presence of phytotoxins that may inhibit germination,[13][14][15] and the possible tie up of nitrogen by incompletely decomposed lignin.[5] It is very common to see blends of 20–30% compost used for transplanting seedlings at cotyledon stage or later.

Composting can destroy pathogens or unwanted seeds. Unwanted living plants (or weeds) can be discouraged by covering with mulch/compost. The "microbial pesticides" in compost may include thermophiles and mesophiles, however certain composting detritivores such as black soldier fly larvae and redworms, also reduce many pathogens. Thermophilic (high-temperature) composting is well known to destroy many seeds and nearly all types of pathogens (exceptions may include prions). The sanitizing qualities of (thermophilic) composting are desirable where there is a high likelihood of pathogens, such as with manure.

7.4   Composting technologies

A homemade compost tumbler

A modern compost bin constructed from plastics

7.4.1 Overview

In addition to the traditional compost pile, various approaches have been developed to handle different composting processes, ingredients, locations, and applications for the composted product.

There is a large number of different composting systems on the market, for example:

- At the household level: Composting toilet, container composting, vermicomposting

- At the industrial composting (large scale): Aerated Static Pile Composting, vermicomposting, windrow composting etc.

7.4.2 Examples

Vermicomposting

Main article: Vermicomposting

Vermicompost is the product or process of composting through the utilization of various species of worms, usually red wigglers, white worms, and earthworms, to create a heterogeneous mixture of decomposing vegetable or food waste (excluding meat, dairy, fats, or oils), bedding materials, and vermicast. Vermicast, also known as worm castings, worm humus or worm manure, is the end-product of the breakdown of organic matter by species of earthworm.[16] Vermicomposting is widely used in North America for on-site institutional processing of food waste, such as in hospitals and

Rotary screen harvested worm castings

shopping malls. This type of composting is sometimes suggested as a feasible indoor home composting method. Vermicomposting has gained popularity in both these industrial and domestic settings because, as compared with conventional composting, it provides a way to compost organic materials more quickly (as defined by a higher rate of carbon-to-nitrogen ratio increase) and to attain products that have lower salinity levels that are therefore more beneficial to plant mediums.[17]

The earthworm species (or **composting worms**) most often used are red wigglers (*Eisenia fetida* or *Eisenia andrei*),

Food waste - after three years

though European nightcrawlers (*Eisenia hortensis* or *Dendrobaena veneta*) could also be used. Red wigglers are recommended by most vermiculture experts, as they have some of the best appetites and breed very quickly. Users refer to European nightcrawlers by a variety of other names, including *dendrobaenas*, *dendras*, Dutch Nightcrawlers, and Belgian nightcrawlers.

Containing water-soluble nutrients, vermicompost is a nutrient-rich organic fertilizer and soil conditioner in a form that is relatively easy for plants to absorb.[18] Worm castings are sometimes used as an organic fertilizer. Because the earthworms grind and uniformly mix minerals in simple forms, plants need only minimal effort to obtain them. The worms' digestive systems also add beneficial microbes to help create a "living" soil environment for plants.

Vermicompost tea in conjunction with 10% castings has been shown to cause up to a 1.7 times growth in plant mass over plants grown without.[19]

Researchers from the Pondicherry University discovered that worm composts can also be used to clean up heavy metals. The researchers found substantial reductions in heavy metals when the worms were released into the garbage and they are effective at removing lead, zinc, cadmium, copper and manganese.[20]

Hügelkultur (raised garden beds or mounds)

Main article: Hügelkultur

The practice of making raised garden beds or mounds filled with rotting wood is also called "Hügelkultur" in German.[21][22] It is in effect creating a Nurse log, however, covered with dirt.

An almost completed Hügelkultur bed (does not have dirt on it yet).

Benefits of hügelkultur garden beds include water retention and warming of soil.[21][23] Buried wood becomes like a sponge as it decomposes, able to capture water and store it for later use by crops planted on top of the hügelkultur bed.[21][24]

The buried decomposing wood will also give off heat, as all compost does, for several years. These effects have been used by Sepp Holzer for one to allow fruit trees to survive at otherwise inhospitable temperatures and altitudes.[22]

Black soldier fly larvae composting

Main article: Hermetia illucens § Uses in composting or as food for animals

Black Soldier Fly (*Hermetia illucens*) larvae have been shown to be able to rapidly consume large amounts of organic waste when kept at 31.8°C, the optimum temperature for reproduction. [25] Enthusiasts have experimented with a large number of different waste products[26] and some even sell starter kits to the public.[27]

Cockroach composting

Cockroach composting is another insect-mediated composting method. In this case the adults of any number of cockroach species (such as the Turkestan cockroach or *Blaptica dubia*) are used to quickly convert manure or kitchen waste to nutrient dense compost. Depending on species used and environmental conditions, excess composting insects can be used as an excellent animal feed for farm animals and pets.[28]

Bokashi

Bokashi is a method that uses a mix of microorganisms to cover food waste to decrease smell. It derives from the practice of Japanese farmers centuries ago of covering food waste with rich, local soil that contained the microorganisms that would ferment the waste. After a few weeks, they would bury the waste. [29]

Most practitioners obtain the microorganisms from the product Effective Microorganisms (EM1),[29] first sold in the 1980s. EM1 is mixed with a carbon base (e.g. sawdust or bran) that it sticks to and a sugar for food (e.g. molasses). The mixture is layered with waste in a sealed container and after a few weeks, removed and buried.[29]

Newspaper fermented in a lactobacillus culture can be substituted for bokashi bran for a successful bokashi bucket. [30]

Compost tea

Compost teas are defined as water extracts brewed from composted materials and can be derived from aerobic or anaerobic processes.[31] Compost teas are generally produced from adding one volume of compost to 4-10 volumes of water, but there has also been debate about the benefits of aerating the mixture.[31] Field studies have shown the benefits of adding compost teas to crops due to the adding of organic matter, increased nutrient availability and increased microbial activity.[31] They have also been shown to have an effect on plant pathogens.[32]

Composting toilets

Main article: Composting toilet

A composting toilet does not require water or electricity, and when properly managed does not smell. A composting toilet collects human excreta which is then added to a compost heap together with sawdust and straw or other carbon rich materials, where pathogens are destroyed to some extent. The amount of pathogen destruction depends on the temperature (mesophilic or thermophilic conditions) and composting time.[33] A composting toilet tries to process the excreta in situ although this is often coupled with a secondary external composting step. The resulting compost product has been given various names, such as humanure and EcoHumus.[33]

Inside a recently started bokashi bin. The aerated base is just visible through the food scraps and bokashi bran.

A composting toilet can aid in the conservation of fresh water by avoiding the usage of potable water required by the typical flush toilet. It further prevents the pollution of ground water by controlling the fecal matter decomposition before entering the system. When properly managed, there should be no ground contamination from leachate.

7.5   Compost and land-filling

As concern about landfill space increases, worldwide interest in recycling by means of composting is growing, since composting is a process for converting decomposable organic materials into useful stable products.[34] Composting is one of the only ways to revitalize soil vitality due to phosphorus depletion in soil.[35] Industrial scale composting in the form of in-vessel composting, aerated static pile composting, and anaerobic digestion takes place in most Western countries now, and in many areas is mandated by law. There are process and product guidelines in Europe that date to the early 1980s (Germany, the Netherlands, Switzerland) and only more recently in the UK and the US. In both these countries, private trade associations within the industry have established loose standards, some say as a stop-gap measure to discourage independent government agencies from establishing tougher consumer-friendly standards.[36] The USA is the only Western country that does not distinguish sludge-source compost from green-composts, and by default in the USA 50% of states expect composts to comply in some manner with the federal EPA 503 rule promulgated in 1984 for sludge products.[37] Compost is regulated in Canada and Australia as well.

A large compost pile that is steaming with the heat generated by thermophilic microorganisms.

7.5.1 Industrial systems

Industrial composting systems are increasingly being installed as a waste management alternative to landfills, along with other advanced waste processing systems. Mechanical sorting of mixed waste streams combined with anaerobic digestion or in-vessel composting is called mechanical biological treatment, and are increasingly being used in developed countries due to regulations controlling the amount of organic matter allowed in landfills. Treating biodegradable waste before it enters a landfill reduces global warming from fugitive methane; untreated waste breaks down anaerobically in a landfill, producing landfill gas that contains methane, a potent greenhouse gas.

Vermicomposting, also known as vermiculture, is used for medium-scale on-site institutional composting, such as for food waste from universities and shopping malls: selected either as a more environmental choice, or to reduce the cost of commercial waste removal.

Large-scale composting systems are used by many urban areas around the world. Co-composting is a technique that combines solid waste with de-watered biosolids, although difficulties controlling inert and plastics contamination from municipal solid waste makes this approach less attractive. The World's largest MSW co-composter is the Edmonton Composting Facility in Edmonton, Alberta, Canada, which turns 220,000 tonnes of residential solid waste and 22,500 dry tonnes of biosolids per year into 80,000 tonnes of compost. The facility is 38,690 meters^2 (416,500 ft^2), equivalent to 4½ Canadian football fields, and the operating structure is the largest stainless steel building in North America, the size of 14 NHL rinks. In 2006, the State of Qatar awarded Keppel Seghers Singapore, a subsidiary of Keppel Corporation to begin construction on a 275,000 tonne/year Anaerobic Digestion and Composting Plant licensed by Kompogas Switzerland. This plant, with 15 independent anaerobic digestors will be the world's largest composting facility once fully operational in early 2011 and forms part of the Qatar Domestic Solid Waste Management Center, the largest integrated waste management complex in the Middle East.

Another large MSW composter is the Lahore Composting Facility in Lahore, Pakistan, which has a capacity to convert 1,000 tonnes of municipal solid waste per day into compost. It also has a capacity to convert substantial portion of the intake into Refuse-derived fuel (RDF) materials for further combustion use in several energy consuming industries across Pakistan e.g., in cement manufacturing companies where it is used to heat up the Cement Kiln systems. This project has also been approved by the Executive Board of the United Nations Framework Convention on Climate Change (UNFCCC) for reduction of emission of methane gas into the climate and has been registered with a capacity of reducing 108,686 metric tonnes CO_2 equivalent per annum.[38]

7.6   Related technologies

Anaerobic digestion is another possible process for converting organic waste into a useful produce (biogas). In central Europe, anaerobic digestion is now more common than composting as a process for treating organic waste. The two processes can also be used in combination: sewage sludge is often anaerobically digested first, followed by a composting process before selling or giving away the compost to farmers.

7.7   History

Composting as a recognized practice dates to at least the early Roman Empire since Pliny the Elder (AD 23-79). Traditionally, composting involved piling organic materials until the next planting season, at which time the materials would have decayed enough to be ready for use in the soil. The advantage of this method is that little working time or effort is required from the composter and it fits in naturally with agricultural practices in temperate climates. Disadvantages (from the modern perspective) are that space is used for a whole year, some nutrients might be leached due to exposure to rainfall, and disease-producing organisms and insects may not be adequately controlled.

Composting was somewhat modernized beginning in the 1920s in Europe as a tool for organic farming. The first industrial station for the transformation of urban organic materials into compost was set up in Wels, Austria in the year 1921.[39] Early frequent citations for propounding composting within farming are for the German-speaking world Rudolf Steiner, founder of a farming method called biodynamics, and Annie Francé-Harrar, who was appointed on behalf of the government in Mexico and supported the country 1950–1958 to set up a large humus organization in the fight against erosion and soil degradation. In the English-speaking world it was Sir Albert Howard who worked extensively in India on sustainable practices and Lady Eve Balfour who was a huge proponent of composting. Composting was imported to America by various followers of these early European movements by the likes of J.I. Rodale (founder of Rodale Organic Gardening), E.E. Pfeiffer (who developed scientific practices in biodynamic farming), Paul Keene (founder of Walnut Acres in Pennsylvania), and Scott and Helen Nearing (who inspired the back-to-the-land movement of the 1960s). Coincidentally, some of the above met briefly in India - all were quite influential in the U.S. from the 1960s into the 1980s.

There are many modern proponents of rapid composting that attempt to correct some of the perceived problems associated with traditional, slow composting. Many advocate that compost can be made in 2 to 3 weeks.[40] Many such short processes involve a few changes to traditional methods, including smaller, more homogenized pieces in the compost, controlling carbon-to-nitrogen ratio (C:N) at 30 to 1 or less, and monitoring the moisture level more carefully. However, none of these parameters differ significantly from the early writings of Howard and Balfour, suggesting that in fact modern composting has not made significant advances over the traditional methods that take a few months to work. For this reason and others, many modern scientists who deal with carbon transformations are sceptical that there is a "super-charged" way to get nature to make compost rapidly.

In fact, both sides are right to some extent. The bacterial activity in rapid high heat methods breaks down the material to the extent that pathogens and seeds are destroyed, and the original feedstock is unrecognizable. At this stage, the compost can be used to prepare fields or other planting areas. However, most professionals recommend that the compost be given time to cure before using in a nursery for starting seeds or growing young plants. The curing time allows fungi to continue the decomposition process and eliminating phytotoxic substances.

Many countries such as Wales[41][42] and some individual cities such as Seattle and San Francisco require food and yard waste to be sorted for composting.[43][44]

Kew Gardens in London has one of the biggest non-commercial compost heaps in Europe.

7.8   See also

- List of composting systems

- San Francisco Mandatory Recycling and Composting Ordinance

- Terra preta

- Urban agriculture

- Waste sorting

7.9   References

[1] "Composting for the Homeowner - University of Illinois Extension". Web.extension.illinois.edu. Retrieved 2013-07-18.

[2] "Composting for the Homeowner - University of Illinois Extension". *uiuc.edu*.

[3] Radovich, T; Hue, N; Pant, A (2011). "Chapter 1: Compost Quality". In Radovich, T; Arancon, N. *Tea Time in the Tropics - a handbook for compost tea production and use*. College of Tropical Agriculture and Human Resources, University of Hawaii. pp. 8–16. External link in |title= (help)

[4] Klickitat County WA, USA Compost Mix Calculator

[5] "The Effect of Lignin on Biodegradability - Cornell Composting". *cornell.edu*.

[6] Dougherty, Mark. (1999). Field Guide to On-Farm Composting. Ithaca, New York: Natural Resource, Agriculture, and Engineering Service.

[7] "Composting - Compost Microorganisms". *Cornell University*. Retrieved 6 October 2010.

[8] Stockholm Environment Institute - EcoSanRes - Guidelines on the Use of Urine and Feces in Crop Production

[9] "TUBdok: Pharmaceutical Residues in Urine and Potential Risks related to Usage as Fertiliser in Agriculture" (PDF). *tu-harburg.de*.

[10] Jenkins, J.C. (2005). *The Humanure Handbook: A Guide to Composting Human Manure*. Grove City, PA: Joseph Jenkins, Inc.; 3rd edition. p. 255. ISBN 978-0-9644258-3-5. Retrieved April 2011.

[11] Courtney Symons (13 October 2011). "'Humanure' dumping sickens homeowner". *YourOttawaRegion*. Metroland Media Group Ltd. Retrieved 16 October 2011.

[12] Healthy Soils, Healthy Landscapes

[13] Morel, P. and Guillemain, G. 2004. Assessment of the possible phytotoxicity of a substrate using an easy and representative biotest. Acta Horticulture 644:417–423

[14] Itävaara et al. Compost maturity - problems associated with testing. in Proceedings of Composting. Innsbruck Austria 18-21.10.2000

[15] Aslam DN, et al. "Development of models for predicting carbon mineralization and associated phytotoxicity in compost-amended soil.". *nih.gov*.

[16] "Paper on Invasive European Worms". Retrieved 22 February 2009.

[17] Lazcano, Cristina; Gómez-Brandón, María; Domínguez, Jorge (2008). "Comparison of the effectiveness of composting and vermicomposting for the biological stabilization of cattle manure" (PDF). *Chemosphere* **72**: 1013–1019. doi:10.1016/j.chemosphere.2008.04.016.

[18] Coyne, Kelly and Erik Knutzen. *The Urban Homestead: Your Guide to Self-Sufficient Living in the Heart of the City*. Port Townsend: Process Self Reliance Series, 2008.

[19] "Worm casting organic fertilizer benefits and uses". *Worms Etc*.

[20] *Cleaning up heavy metals using worms*, International: mining.com, 2012, retrieved 3 October 2012

[21] "hugelkultur: the ultimate raised garden beds". Richsoil.com. 2007-07-27. Retrieved 2013-07-18.

[22] "The Art and Science of Making a Hugelkultur Bed - Transforming Woody Debris into a Garden Resource Permaculture Research Institute - Permaculture Forums, Courses, Information & News". Retrieved 2013-07-18.

[23] "Hugelkultur: Composting Whole Trees With Ease Permaculture Research Institute - Permaculture Forums, Courses, Information & News". Retrieved 2013-07-18.

[24] Hemenway, Toby (2009). Gaia's Garden: A Guide to Home-Scale Permaculture. Chelsea Green Publishing. pp. 84-85. ISBN 978-1-60358-029-8.

[25] Diener, Stefan; Studt Solano, Nandayure M.; Roa Gutiérrez, Floria; Zurbrügg, Christian; Tockner, Klement (2011). "Biological Treatment of Municipal Organic Waste using Black Soldier Fly Larvae". *Waste and Biomass Valorization* **2** (4): 357–363. doi:10.1007/s12649-011-9079-1. ISSN 1877-2641.

[26] "E". *Bio-Conversion of Putrescent Waste*. ESR International. Retrieved 17 April 2015.

[27] "BSF Farming - marketplace". Retrieved 17 April 2015.

[28] "Cockroach Composting". *The Unconventional Farmer*.

[29] Lindsay, Jay (12 June 2012). "Japanese composting may be new food waste solution". *AP*. Retrieved 13 November 2012.

[30] "Make your own FREE bokashi starter", 12 September 2008. Retrieved 7 November 2013.

[31] Gómez-Brandón, M; Vela, M; Martinez Toledo, MV; Insam, H; Domínguez, J (2015). "12: Effects of Compost and Vermicompost Teas as Organic Fertilizers". In Sinha, S; Plant, KK; Bajpai, S. *Advances in Fertilizer Technology: Synthesis (Vol1)*. Stadium Press LLC. pp. 300–318. ISBN 1-62699-044-1.

[32] Santos, M; Dianez, F; Carretero,F (2011). "12: Suppressive Effects of Compost Tea on Phytopathogens". In Dubey,NK. *Natural products in plant pest management*. Oxfordshire, UK Cambridge, MA: CABI. pp. 242–262. ISBN 9781845936716.

[33] Stenström, T.A., Seidu, R., Ekane, N., Zurbrügg, C. (2011). Microbial exposure and health assessments in sanitation technologies and systems - EcoSanRes Series, 2011-1. Stockholm Environment Institute (SEI), Stockholm, Sweden, page 88

[34] A Brief History of Solid Waste Management

[35] "Preventing Contaminants in Home Compost Piles". Retrieved 16 June 2012.

[36] "US Composting Council". Compostingcouncil.org. Retrieved 2013-07-18.

[37] "Electronic Code of Federal Regulations. Title 40, part 503. Standards for the use or disposal of sewage sludge". *U.S. Government Printing Office*. 1998. Retrieved 30 March 2009.

[38] Details on project design and its validation and monitoring reports are available at: Project 2778 : Composting of Organic Content of Municipal Solid Waste in Lahore

[39] Welser Anzeiger vom 05. Januar 1921, 67. Jahrgang, Nr. 2, S. 4

[40] The Rapid Compost Method by Robert Raabe, Professor of Plant Pathology, Berkeley

[41] Gwynedd Council food recycling

[42] "Anglesey households achieve 100% food waste recycling". *edie.net*.

[43] "San Francisco Signs Mandatory Recycling & Composting Laws". Retrieved 19 September 2010.

[44] Tyler, Aubin (21 March 2010). "The case for mandatory composting". *The Boston Globe*. Retrieved 19 September 2010.

Chapter 8

Cottonseed meal

Cottonseed meal is the byproduct remaining after cotton is ginned and the seeds crushed and the oil extracted. The remaining meal is usually used for animal feed and in organic fertilizers.[1] However, the meal can be fed only to adult ruminants because it contains a compound called gossypol. The compound is highly toxic to monogastrics and even sometimes to calves which have less well-developed digestive systems.[2]

8.1 References

[1] Card, Adrian; David Whiting; Carl Wilson; Jean Reeder (2009). "CMG Garden Notes #234 Organic Fertilizers" (PDF). Colorado State University Extension. p. 3. Retrieved 19 April 2011.

[2] Morgan, Sandra. "Gossypol Toxicity in Livestock" (PDF). Oklahoma State University. Retrieved 19 April 2011.

Chapter 9

Effluent spreading

Effluent spreading is a process in which a slurry of effluent from a dairy farm's milking parlor is pumped and spread on pasture. Commonly a rotating sprinkler is used. Dairy manure contains ammonium NH4-N.

In New Zealand the application of effluent is a permitted activity, although spreading in excess is an environmental hazard.

Chapter 10

Feather meal

Feather meal is a byproduct of processing poultry; it is made from poultry feathers by partially hydrolyzing them under elevated heat and pressure, and then grinding and drying. Although total nitrogen levels are fairly high (up to 12%), the bioavailability of this nitrogen may be low. Feather meal is used in formulated animal feed and in organic fertilizer.

Worldwide, more than 25 billion chickens are used for human consumption. Feather meal is made through a process called rendering. Steam pressure cookers with temperatures over 140°C are used to "cook" and sterilize the feathers. This partially hydrolyzes the proteins, which denatures them. It is then dried, cooled and ground into a powder for use as a nitrogen source for animal feed (mostly ruminants) or as an organic soil amendment.

Containing up to 12% nitrogen, it is a source of slow-release, organic, high-nitrogen fertilizer for organic gardens. It is not water-soluble and does not make a good liquid fertilizer. It can be used to:

- Increase green leaf growth

- Activate compost decomposition

- Improve soil structure

When adding it to a garden as a nitrogen source, it must be blended into the soil to start the decomposition to make the nitrogenous compounds available to the plants. As an organic garden fertilizer, it is not synthetic or petroleum-based.

10.1 Sources

- Organic gardening information

- Waste not, want not?Poultry "feather meal" as another source of antibiotics in feed // Tara C. Smith, April 5, 2012

Chapter 11

Fish emulsion

Fish emulsion is a fertilizer emulsion that is produced from the fluid remains of fish processed for fish oil and fish meal industrially.

11.1 Gardening

Since fish emulsion is naturally derived, it is considered an organic fertilizer appropriate for use in organic horticulture. In addition to having a typical N-P-K analysis of 5-2-2, fish emulsion adds micronutrients.[1]

11.2 See also

- Fish hydrolysate
- Organic fertilizers

11.3 References

[1] Colorado State University - Cooperative Extension. "Organic Fertilizers." GardenNotes #234.

Chapter 12

Fish hydrolysate

Fish hydrolysate, in its simplest form, is ground up fish into a liquid phase where cleavage of natural bonds can be enacted through various biological processes. Raw material choice of whole fish or by products is made pending the commercial outlets for the fish. In some cases, the fillet portions are removed for human consumption, the remaining fish body, which means the guts, bones, cartilage, scales, meat, etc., is put into water and ground up. Some fish hydrolysate is ground more finely than others so more bone material is able to remain suspended. Enzymes may also be used to solubilize bones, scale and meat. If the larger chunks of bone and scales are screened out, calcium or protein, or mineral content may be lacking in the finished product form. If purchasing fish hydrolysate for agricultural applications, one should look at the label carefully for the concentration of mineral elements in the liquid. Some fish hydrolysates have been made into a dried product, increasing the potential food/feed ingredient inclusion opportunities. The oil is separated out in this process, which means the Omega 3 functional food component would be remain with the oil and not the hydrolysate.[1]

The nutritional benefits of hydrolyzing fish has spawned a new industry producing on fish protein powder for Food and Nutritional applications which aims to capitalize on the value of the fish peptides produced as a result of enzymatic action on the fish protein.

12.1   Uses of fish hydrolysate

There is a lot of information on the many uses of fish hydrolysate - from its low end use as a fish-based fertilizer, to its use as an animal food, or new human consumption applications which are developing. A number of scientific journals have cited the antiproliferative activity of fish protein hydrolysate, which makes it eligible for listing as a nutriceutical. Fish protein hydrolysates, particularly those developed from salmon, contain significant cancer growth inhibitors.[2] and those from sardines have gut health and anti-hypertension benefits.

12.2   Bycatch

New technologies that have increased fishing efficiency have also resulted in the taking of species or sizes not suitable for market, known as bycatch. An increased catch of unsaleable whole fish has resulted from the increased bycatch of the fishing industry. These fish are often dumped overboard at sea, but are also brought into port in the holds of fishing boats. This has spurred an incentive to find a market for the bycatch in order to lower the cost of production.

12.3   Stabilizers

The liquid fish hydrolysate process minces the whole fish, then enzymatically digests, then grinds and liquifies the resulting product, known as gurry. Because it is a cold process, gurry putrefies more rapidly than fish emulsion and needs to be stabilized at a lower pH, requiring more acid. Researchers have tried formic acid, sulfuric acid, and others. Formic acid had phytotoxic effects on plants. Phosphoric acid is the preferred stabilizer.[3]

12.4 Comparison with fish emulsion

If fish hydrolysate is heated, the oils and certain proteins can be more easily removed to be sold in purified forms. The complex protein, carbohydrate and fats in the fish material are denatured, which means they are broken down into less complex foods. Overheating can result in destruction of the material as a food to grow beneficial organisms. Once the oils are removed and proteins denatured and simplified by the heating process, this material is called a fish emulsion. The hydrolysate process has substantially lower capital and production costs compared to fish emulsion production.[1]

12.5 References

[1] Critical Reviews in Food Science and Nutrition, Volume 40, Issue 1 January 2000, pp. 43-81.

[2] Picot, L, et al., 'Antiproliferative activity of fish protein hydrolysates on human breast cancer cell lines', Process Biochemistry. Vol. 41, no. 5, May 2006.

[3] Brian Baker, *Plant Nutrition from the Sea*, Farmer to Farmer 16 (Sept./Oct. 1996), available online at .

12.6 External links

- Fish Protein Hydrolysates: Production, Biochemical, and Functional Properties

- Freshwater Fish Processing - Equipment and Examples of Technological Lines

- Alaska Bounty - Plant Food From the Sea

12.7 Further reading

- Hordur G. Kristinsson & Barabar A. Rasco, *Fish Protein Hydrolysates: Production, Biochemical, and Functional Properties* pp. 43–81, 40 Institute for Food Science and Technology 1(University of Washington 2000).

Chapter 13

Fish meal

Fish meal factory, Bressay

Fish meal, or **fishmeal**, is a commercial product made from fish and the bones and offal from processed fish. It is a brown powder or cake obtained by drying the fish or fish trimmings, often after cooking, and then grinding it. If it is a fatty fish it is also pressed to extract most of the fish oil.[1]

13.1   History

Fish byproducts have been used historically to feed poultry, pigs, and other farmed fish. A primitive form of fishmeal is mentioned in *The Travels of Marco Polo* at the beginning of the 14th century: 'they accustom their cattle, cows, sheep, camels, and horses to feed upon dried fish, which being regularly served to them, they eat without any sign of dislike.' The use of herring as an industrial raw material started as early as about 800 AD in Norway; a very primitive process of pressing the oil out of herring by means of wooden boards and stones was employed.[2]

13.2   Raw materials used

Fishmeal can be made from almost any type of seafood, but is generally manufactured from wild-caught, small marine fish that contain a high percentage of bones and oil, and is usually deemed not suitable for direct human consumption. The fish caught for fishmeal purposes solely are termed "industrial".[3] Other sources of fishmeal are from bycatch of other fisheries and byproducts of trimmings made during processing (fish waste or offal) of various seafood products destined for direct human consumption. Virtually any fish or shellfish in the sea can be used to make fishmeal, although a few rare unexploited species may produce a poisonous meal.

13.3   Selecting species

In selecting which species to use:

1. The species must be in large concentrations to give a high catching rate; this is essential because the value of industrial fish is less than that of fish for direct human consumption.

2. The fishery should preferably be based on more than one species to reduce the effect of fluctuations in supply of any one species.

3. The total abundance of long-lived species varies less from year to year

4. Species with a high fat content are more profitable, because the fat in fish is held at the expense of water and not at the expense of protein.

Fish meal is manufactured primarily from anchovies in Peru, menhaden in the United States, pout in Norway, capelin, sand eel, and mackerel in other parts of northern Europe, and sauries, mackerels, and sardines in Japan.[1]

13.4   Production

Fishmeal is made by cooking, pressing, drying, and grinding of fish or fish waste to which no other matter has been added. It is a solid product from which most of the water is removed and some or all of the oil is removed. Four or five tonnes of fish are needed to manufacture one tonne of dry fishmeal.[4]

Of the several ways of making fishmeal from raw fish, the simplest is to let the fish dry out in the sun. This method is still used in some parts of the world where processing plants are not available, but the end product is poor quality in comparison with ones made by modern methods. Now, all industrial fish meal is made by the following processes:[2]

Cooking: A commercial cooker is a long, steam-jacketed cylinder through which the fish are moved by a screw conveyor. This is a critical stage in preparing the fishmeal, as incomplete cooking means the liquid from the fish cannot be pressed out satisfactorily and overcooking makes the material too soft for pressing. No drying occurs in the cooking stage.

Pressing: A perforated tube with increasing pressure is used for this process. This stage involves removing some of the oil and water from the material and the solid is known as press cake. The water content in pressing is reduced from 70% to about 50% and oil is reduced to 4%.

Fish meal factory, Bressay, Shetland Islands

Drying: If the meal is under-dried, moulds or bacteria may grow. If it is over-dried, scorching may occur and this reduces the nutritional value of the meal.

The two main types of dryers are:

- Direct: Very hot air at a temperature of $500°C$ ($932°F$) is passed over the material as it is tumbled rapidly in a cylindrical drum. This is the quicker method, but heat damage is much more likely if the process is not carefully controlled.

- Indirect: A cylinder containing steam-heated discs is used, which also tumbles the meal.

Grinding: This last step in processing involves the breakdown of any lumps or particles of bone.

13.5 Nutrient composition

Any complete diet must contain some protein, but the nutritional value of the protein relates directly to its amino acid composition and digestibility. The amino acid profile of fishmeal makes this feed ingredient attractive as a protein supplement. High-quality fishmeal normally contains between 60% and 72% crude protein by weight. Typical diets for fish may contain from 32% to 45% total protein by weight.[5] Fishmeal is sought after as an ingredient in aquaculture diets because it contains compounds that make the feed more palatable. This allows the feed to be ingested rapidly, and will reduce nutrient leaching. Nonessential glutamic acid is an amino acid thought to impart palatability to fishmeal.[6]

Fish lipids are also highly digestible by all species of animals and are excellent sources of the essential polyunsaturated fatty acids (PUFA), including both omega-3 and omega-6 fatty acids. The predominant omega-3 fatty acids in fishmeal and fish oil are linolenic acid, docosahexaenoic acid, and eicosapentaenoic acid. Essential fatty acids are necessary for normal larval development, fish growth, and reproduction. They are important in normal development of the skin, nervous

Fish meal factory, Westfield, West Lothian

system, brain, and visual acuity. PUFAs appear to assist the immune system in defense of disease agents and reduce the stress response. Fishmeal also contains valuable phospholipids, fat-soluble vitamins, and steroid hormones.[7]

High digestibility of fish lipids means they can provide considerable usable energy. If a diet does not provide enough energy, the fish or shrimp will have to break down valuable protein for energy, which is expensive and can increase production of toxic ammonia. Fishmeal is considered to be a moderately rich source of vitamins of the B-complex, especially cyanocobalamine (B_{12}), niacin, choline, pantothenic acid, and riboflavin.

13.6 Benefits

Fishmeal in diets increases feed efficiency and growth through better feed palatability, and enhances nutrient uptake, digestion, and absorption. The balanced amino acid composition of fishmeal complements and provides synergistic effects with other animal and vegetable proteins in the diet to promote fast growth and reduce feeding costs.

High-quality fishmeal provides a balanced amount of all essential amino acids, phospholipids, and fatty acids required for optimum development, growth, and reproduction, especially of larvae and broodstock. The nutrients in fishmeal also aid in disease resistance by boosting and helping to maintain a healthy functional immune system. It also allows for formulation of nutrient-dense diets, which promote optimal growth.[8]

Incorporation of fishmeal into diets of aquatic animals helps to reduce pollution from the wastewater effluent by providing greater nutrient digestibility. The incorporation of high-quality fishmeal into feed imparts a 'natural or wholesome' characteristic to the final product, such as that provided by wild fish.

13.7 In pet food

Fish meal is found in approximately 8% of pet foods as a non-descriptive source of protein, low in omega fatty acids as the oil is pressed out. The quality of the ingredient in pet food is suspect, as poor quality and rancid fish are often used.[9]

13.8 Ecological links

Aquaculture's heavy reliance on wild-caught seed and broodstock is of increasing concern. Fishmeal and its source of raw materials and costs are highly debated by scientists and conservationists. Since fishmeal uses wild fish stock to feed farmed fish, this places direct pressure on fisheries resources. Indirect effects are also apparent such as diminishing wild fisheries, habitat modification and food web interactions. The trace contaminants in the feed, if present, can cause diseases and fish mortality.

13.9 Economy

About 23.13 million tonnes of compound aquafeeds were produced in 2005, of which about 42% was consumed by aquaculture. The aquaculture sector consumed around 3.06 million tonnes or 56% of world fishmeal production and 0.78 million tonnes or 87% of total fish oil production in 2006, with over 50%of fish oil going into salmonid diets. Increasing prices of fishmeal, fish oil, grains and other feed ingredients, and also fuel and energy will certainly affect the cost of aquaculture production.[10] Sustainability remains a concern, however, even more so when the demand for aquaculture products is outstripping the supply, and prices soar so that even inefficient farms might make money.

13.10 Risks

Unmodified fish meal can spontaneously combust. In the past, ships have sunk because of such fires. Now, the danger is eliminated by adding antioxidants, namely, ethoxyquin. There has been some speculation that ethoxyquin in pet foods might be responsible for multiple health problems. To date, the U.S. Food and Drug Administration has only found a verifiable connection between ethoxyquin and buildup of protoporphyrin IX in the liver, as well as elevations in liver-related enzymes in some animals, but with no known health consequences from these effects. In 1997, the Center for Veterinary Medicine asked pet food manufacturers to voluntarily limit ethoxyquin levels to 75 ppm until further evidence is reported. However, most pet foods that contain ethoxyquin have never exceeded this amount. Ethoxyquin has been shown to be slightly toxic to fish.

Ethoxyquin is not permitted for use in Australian foods, nor is it approved for use within in the European Union, though it is an accepted additive in the U.S. Besides the USA, it is also widely used in other third-world countries.

Though it has been approved for use in foods in the US, and as a spray insecticide for fruits, ethoxyquin has surprisingly not been thoroughly tested for its carcinogenic potential. Ethoxyquin has long been suggested to be a possible carcinogen, and a very closely related chemical, 1,2-dihydro-2,2,4-trimethylquinoline, has been shown to have carcinogenic activity in rats, and a potential for carcinogenic effect to fishmeal prior to storage or transportation.[4]

13.11 Future

Despite the adverse effects, organisations such as the Fishmeal Information Network (FIN), which is one source of contact for fishmeal and gives information on its supply chain and its role in the nutrition of farm livestock. FIN aims to present fact-based information, independent evidence, and respected expert opinion on fishmeal and its use. It recognises the imperatives of safety in the food chain, healthy diets, animal welfare, and protection of the environment.

The FIN monitors two key areas: Legislation, which governs fishmeal use in animal feed, and contaminant issues and regulations that have, or are likely to have, an impact on fishmeal, fish oil, wild finfish and farmed fish. Such regulations and precautions will help companies better their products and will benefit the consumers greatly. More research into this area is needed to make effective decisions and to obtain optimal results.

13.12   Main producing countries

- Chile: anchovy, horse mackerel

- China: various species

- Denmark: pout, sandeel, sprat

- European Union: various species

- Iceland and Norway: capelin, herring, blue whiting

- Japan: sardine, pilchard

- Peru: anchovy

- South Africa: pilchard

- Thailand: various species

- United States: menhaden, pollock

13.13   References

[1] Pauly, Daniel and Watson, Reg (2009) "Spatial Dynamics of Marine Fisheries" In: Simon A. Levin (ed.) *The Princeton Guide to Ecology*. Pages 501–509.

[2] Windsor, M L. (2001). Fish Meal. Department of Trade and Industry Torry Research. TORRY ADVISORY NOTE No. 49

[3] The fish site

[4] Miles RD and Chapman FA (2006) "The Benefits of Fish Meal in Aquaculture Diets" *University of Florida*. Document FA122, p.6.

[5] "Manufacture, Storage, Composition And Use Of Fish Meal".

[6] Johnston, I. A., S. Manthri, et al. (2002). "Effects of dietary protein level on muscle cellularity and flesh quality in Atlantic salmon with particular reference to gaping." *Aquaculture* 210(1-4): 259–283.

[7] Regost, C., J. Arzel, et al. (2001). "Fat deposition and flesh quality in seawater reared, triploid brown trout (Salmo trutta) as affected by dietary fat levels and starvation." *Aquaculture* 193(3-4): 325–345.

[8] Lie, Ø. (2001). "Flesh quality – the role of nutrition." *Aquaculture Research* 32: 341–348.

[9] "Pet food nutrition and ingredient analysis".

[10] "Higher fishmeal prices result in good business".

Chapter 14

Gomutra

Gomutra (Sanskrit: गोमूत्र; lit. *cow urine*) refers to the usage of cow urine for therapautic purposes in traditional Indian medicine, Ayurveda.[1] It also used in Vaastu Shastra for purification purposes.[2]

Gomutra is also an important component of the mixture called Panchagavya also used in Ayurveda.[1] Urine of a pregnant cow is considered special and it is claimed that it contains special hormones and minerals.[2]

14.1 Claimed benefits and usage

14.1.1 In religious rituals

Sprinkling of cow urine is said to have a spiritual cleansing effect.[3][4]

14.1.2 For pharmaceutical purposes

In Ayurveda, Gomutra is claimed to be helpful in the treatment of leprosy and cancer. A mixture of gomutra, Triphala, and cow milk is used for the treatment of Anaemia. It is also used in the treatment of fever by mixing it with black pepper, yoghurt, and ghee (*ghrita*). A mixture of gomutra, neem bark, vasaka bark, kurilo bark, kaner leaves. A mixture of gomutra and dharuharidra is used for epilepsy.[1] In study from Mandsaur has claimed that it may also benefit cancer patients.[5]

Cow urine is also used in Myanmar and Nigeria as a folk medicine.[6][7] In Nigeria, a concoction of leaves of tobacco, garlic and lemon basil juice, rock salt and cow urine is used to treat convulsions in children.[7] This has resulted in the death of several children from respiratory depression.[8]

According to the head of the Ayurvedic institute Dhanwanthari Vaidyasala of Thodupuzha, Satish Namboodiri, it is also used for peptic ulcer, certain type of cancer, liver ailments, and asthma.[9]

In 2002, a US patent was issued to a group of Indian scientists from Council of Scientific and Industrial Research (CSIR). It was an antibiotic and cow urine distillate mixture. The cow urine was claimed to be serving as a bioenhancer, enhancing anti-microbial activity of antibiotic and antifungal agents.[10][11][12]

In 2010, Rashtriya Swayamsevak Sangh-funded Go-vigyan Anusandhan Kendra in Deolapar and National Environmental Engineering Research Institute (NEERI) acquired an US patent on a gomutra-based drug. They claimed that the patent validates that gomutra has anti-cancer properties. The patent was for a gomutra-based mixture that claimed to prevent oxidative damage to DNA.[13][14]

Gomutra is also marketed as a health drink. In 2009, Kanpur Gaushala Society (KGS) in Kanpur released a product called Goloka Pay, a cold drink containing 5% distilled cow urine. It was released in two flavours, orange and lemon. It also contained herbs like tulsi, shankhpushpi and brahmi.[15] Also in 2009, the Cow Protection Department of Rashtriya

Swayamsevak Sangh in Haridwar announced their plans to release a similar product.[16]

Comsetic products like soaps and shampoos are also made from gomutra.[17][18]

14.1.3   As a floor cleaner

A floor-cleaning fluid called Gaunyle is marketed by an organisation called Holy Cow Foundation.[19] Maneka Gandhi, Women and Child Development Minister, has proposed than Gaunyle be used instead of Phenyl in government offices.[20] In May 2015, Rajendra Singh Rathore, Medical and Health Minister of Rajasthan, inaugurated a ₹4,00,00,000 (US$630,000) cow-urine refinery in Jalore.[21][22] The refinery was set up by Parthvimeda Gau Pharma Pvt. Ltd. which produces a floor cleaner called Gocleaner.[22]

14.1.4   In organic farming

Jeevamrutha *storage cans*

Gomutra is used as a manure for production of rice.[23] *Jeevamrutha* is a fertilizer made from a mixture of cow urine, cow dung, jaggery, pulse flour and rhizosphere soil.[24] A mixture of gomutra, custard apple leaves and neem leaves after boiling is said to serve as a biopesticide.[23]

In 2012, College of Veterinary & Animal Sciences in Wayanad district began selling packaged gomutra and Panchagavya. The products were primary directed towards organic farming with claims that it would reduce usage of chemical fertilizers and pesticides. Gomutra is supposed to increase plant resistance and Panchagavya is supposed to increase growth of soil bacteria and improve fertility. The head of the institute, Joseph Mathew, said that quality was assured by collecting by the first urine of the day from the cows. He added that it cannot be used of medicinal purposes.[9]

14.2   Scientific studies

A 1975 study on mice found that cow urine causes death in high doses.[25] A similar 1976 study on dogs showed that repeated administration of cow urine concoction as used in Nigerian folk medicine, resulted in hypotension and tachypnea, and also death.[26] A 2001 study found prions in detectable amount in the urine of cows suffering from bovine spongiform encephalopathy.[27]

14.3 See also

- Panchagavya

- Aqua omnium florum

- Alternative cancer treatments

14.4 References

[1] T V Sairam (16 January 2008). *The Penguin Dictionary of Alternative Medicine*. Penguin Books Limited. p. 316. ISBN 978-93-5118-127-9. Retrieved 6 January 2015.

[2] N. H. Sahasrabudhe; R. D. Mahatme (2000). *Mystic Science of Vastu*. Sterling Publishers Pvt. Ltd. p. 68. ISBN 978-81-207-2206-4. Retrieved 6 January 2015.

[3] "Kamadhenu Sutra". *Outlook India*. 10 March 2003. Retrieved 6 January 2015.

[4] "Teachers "purify" students with cow urine". *Reuters*. 23 April 2007. Retrieved 6 January 2015.

[5] N. K. Jain; V. B. Gupta; Rajesh Garg; N. Silawat (2010). "Efficacy of cow urine therapy on various cancer patients in Mandsaur District, India - A survey". *International Journal of Green Pharmacy* **4** (1): 29–35. Retrieved 9 May 2015.

[6] "An amazing cow's urine therapy practice in Myanmar". University of Toyama. Retrieved 29 March 2015.

[7] "Effects of cow urine concoction and nicotine on the nerve-muscle preparation in common African toad Bufo regularis". *Biomedical Research* **16** (16 (3)): 205–211. 2005.

[8] "Don't use cow urine to treat infant epilepsy, Kwara warns mothers". *Premium Times*. 2 February 2013. Retrieved 29 March 2015.

[9] "Cow urine aids treatment of cancer, asthma?". *The Economic Times*. 12 July 2012. Retrieved 6 January 2015.

[10] US 6410059 "Pharmaceutical composition containing cow urine distillate and an antibiotic"

[11] "Indian patents cow urine for medicinal use". *The Indian Express*. 3 July 2002. Retrieved 6 January 2015.

[12] "Cow urine therapy". *The Hindu*. 19 September 2009. Retrieved 6 January 2015.

[13] "RSS-sponsored cow urine drug gets US, China patents". *The Indian Express*. 19 July 2010. Retrieved 6 January 2015.

[14] US 7718360 "Composition (RUCD) for protecting and/or repairing DNA from oxidative damages and a method thereof"

[15] "Pure Cow-Ka Cola". *Outlook India*. 17 August 2009. Retrieved 6 January 2015.

[16] "Coke has a rival: RSS's cow urine cola". *The Indian Express*. 10 February 2009. Retrieved 6 January 2015.

[17] "Holy cow! Vishwa Hindu Parishad launches 'Gau mutra' cosmetics". *Deccan Chronicle*. 10 December 2014. Retrieved 6 January 2015.

[18] "Soaps, shampoos from cow urine!". *DNA India*. 3 October 2011. Retrieved 6 January 2015.

[19] "Use cow urine to clean offices, says Maneka Gandhi". *The Times of India*. 25 March 2015. Retrieved 6 April 2015.

[20] "Cow urine cleaner to replace phenyl in government offices". *India Today*. 9 January 2015. Retrieved 9 May 2015.

[21] "Cow-urine refinery inaugurated at Jalore". *Deccan Herald*. 3 May 2015. Retrieved 9 May 2015.

[22] "Cow urine to be used to clean Rajasthan government hospitals". *India Today*. 5 May 2015. Retrieved 9 May 2015.

[23] "Farmer cultivates paddy with cow urine, dung". *The Hindu*. 13 December 2012. Retrieved 6 January 2015.

[24] T. Satyanarayana; Bhavdish Narain Johri; Anil Prakash (2 January 2012). *Microorganisms in Sustainable Agriculture and Biotechnology*. Springer Science & Business Media. p. 63. ISBN 978-94-007-2214-9. Retrieved 6 January 2015.

[25] DD Oyebola; RA Elegbe (1975). "Cow's urine poisoning in Nigeria. Experimental observations in mice.". *Tropical and geographical medicine* **27** (2): 194–202. Retrieved 29 March 2015.

[26] R. A. Elegbe; D. D. O. Oyebola (1977). "Cow's urine poisoning in Nigeria: cardiorespiratory effects of cow's urine in dogs". *Transactions of the Royal Society of Tropical Medicine and Hygiene* **71** (2): 127–132. Retrieved 29 March 2015.

[27] GM Shaked; Y Shaked; Z Kariv-Inbal; M Halimi (2001). "A protease-resistant prion protein isoform is present in urine of animals and humans affected with prion diseases". *Journal of Biological Chemistry* **276** (34): 31479–31482. Retrieved 29 March 2015.

14.5   External links

- Holy Cow Foundation, an organisation involved in cow protection and promoting various bovine products

sds

Chapter 15

Liquid manure

A liquid manure storage silo, located in the Netherlands.

Liquid manure is manure in a liquid form. Manure is changed into a liquid form by mixing the manure with water. Generally, liquid manure is used as a convenient alternative to manure, which cannot be spread as evenly as its liquid form. Manure in both states is used as a nutrient-enriched fertilizer for plants.<ref name=Farmer's Handbook: Liquid Manure>"Farmer's Handbook: Liquid Manure" (PDF). Permaculture News. Retrieved 2015-07-10.</ref>

15.1 See also

- Chicken manure
- Cow manure
- Manure spreader

15.2 References

15.3 External links

Media related to Liquid manure at Wikimedia Commons

Chapter 16

Manure

This article is about organic material used as soil fertilizer. For animal dung used for other purposes, see feces. "Animal waste" redirects here. For other types of animal waste, see metabolic waste.

Manure is organic matter, mostly derived from animal feces except in the case of green manure, which can be used

Animal manure is often a mixture of animal feces and bedding straw, as in this example from a stable

as organic fertilizer in agriculture. Manures contribute to the fertility of the soil by adding organic matter and nutrients, such as nitrogen, that are trapped by bacteria in the soil. Higher organisms then feed on the fungi and bacteria in a chain of life that comprises the soil food web. It is also a product obtained after decomposition of organic matter like cow dung which replenishes the soil with essential elements and add humus to the soil.

In the past, the term "manure" included inorganic fertilizers, but this usage is now very rare.

16.1   Types

There are three main classes of manures used in soil management:

16.1.1   Animal manure

Cement reservoirs, one new, and one containing cow manure mixed with water. This is common in rural Hainan Province, China.

Most animal manure consist of feces. Common forms of animal manure include farmyard manure (FYM) or farm slurry (liquid manure). FYM also contains plant material (often straw), which has been used as bedding for animals and has absorbed the feces and urine. Agricultural manure in liquid form, known as slurry, is produced by more intensive livestock rearing systems where concrete or slats are used, instead of straw bedding. Manure from different animals has different qualities and requires different application rates when used as fertilizer. For example horses, cattle, pigs, sheep, chickens, turkeys, rabbits, and guano from seabirds and bats all have different properties.[1] For instance, sheep manure is high in nitrogen and potash, while pig manure is relatively low in both. Horses mainly eat grass and a few weeds so horse manure can contain grass and weed seeds, as horses do not digest seeds the way that cattle do. Chicken litter, coming from a bird, is very concentrated in nitrogen and phosphate and is prized for both properties.

Animal manures may be adulterated or contaminated with other animal products, such as wool (shoddy and other hair), feathers, blood, and bone. Livestock feed can be mixed with the manure due to spillage. For example, chickens are often fed meat and bone meal, an animal product, which can end up becoming mixed with chicken litter.

Human manure

Main article: Reuse of excreta

Some people refer to human excreta as human manure, and the word "humanure" has also been used. Just like animal manure, it can be applied as a soil conditioner (reuse of excreta in agriculture). Sewage sludge is a material that contains human excreta, as it is generated after mixing excreta with water and treatment of the wastewater in a sewage treatment plant.

Compost containing turkey manure and wood chips from bedding material is dried and then applied to pastures for fertilizer.

16.1.2 Compost

Main article: Compost

Compost is the decomposed remnants of organic materials. It is usually of plant origin, but often includes some animal dung or bedding.

16.1.3 Green manure

Green manures are crops grown for the express purpose of plowing them in, thus increasing fertility through the incorporation of nutrients and organic matter into the soil. Leguminous plants such as clover are often used for this, as they fix nitrogen using *Rhizobia* bacteria in specialized nodes in the root structure.

Other types of plant matter used as manure include the contents of the rumens of slaughtered ruminants, spent grain (left over from brewing beer) and seaweed.

16.2   Uses of manure

16.2.1   Animal manure

Animal manure, such as chicken manure and cow dung, has been used for centuries as a fertilizer for farming. It can improve the soil structure (aggregation) so that the soil holds more nutrients and water, and therefore becomes more fertile. Animal manure also encourages soil microbial activity which promotes the soil's trace mineral supply, improving plant nutrition. It also contains some nitrogen and other nutrients that assist the growth of plants.

Manures with a particularly unpleasant odor (such as slurries from intensive pig farming) are usually knifed (injected) directly into the soil to reduce release of the odor. Manure from pigs and cattle is usually spread on fields using a manure spreader. Due to the relatively lower level of proteins in vegetable matter, herbivore manure has a milder smell than the dung of carnivores or omnivores. However, herbivore slurry that has undergone anaerobic fermentation may develop more unpleasant odors, and this can be a problem in some agricultural regions. Poultry droppings are harmful to plants when fresh but, after a period of composting, are valuable fertilizers.

Manure is also commercially composted and bagged and sold retail as a soil amendment.

Before motor vehicles became common, horse droppings were a big part of the rubbish that communities needed to clean off roads.

16.3   Precautions

Manure generates heat as it decomposes, and it is possible for manure to ignite spontaneously if stored in a very large pile.[2] Once such a large pile of manure is burning, it will foul the air over a wide area and require considerable effort to extinguish. Therefore, large feedlots must take care to ensure that piles of fresh manure do not get excessively large. There is no serious risk of spontaneous combustion in smaller operations.

There is also a risk of insects carrying feces to food and water supplies, making them unsuitable for human consumption.

16.3.1   Livestock antibiotics

In 2007, a University of Minnesota study[3] indicated that foods such as corn, lettuce, and potatoes have been found to accumulate antibiotics from soils spread with animal manure that contains these drugs.

Organic foods may be much more or much less likely to contain antibiotics, depending on their sources and treatment of manure. For instance, by Soil Association Standard 4.7.38, most organic arable farmers either have their own supply of manure (which would, therefore, not normally contain drug residues) or else rely on green manure crops for the extra fertility (if any nonorganic manure is used by organic farmers, then it usually has to be rotted or composted to degrade any residues of drugs and eliminate any pathogenic bacteria — Standard 4.7.38, Soil Association organic farming standards). On the other hand, as found in the University of Minnesota study, the non-usage of artificial fertilizers, and resulting exclusive use of manure as fertilizer, by organic farmers can result in significantly greater accumulations of antibiotics in organic foods.[3]

16.4   See also

16.5   Notes

[1] "Manure". Bbc.co.uk. Retrieved 2012-11-14.

[2] "Spontaneous Combustion of Manure Starts 200-Acre Blaze 1/08/07 |". *abc7.com*. Retrieved 2010-08-07.

[3] staff (2007-07-12). "Livestock Antibiotics Can End Up in Human Foods". Ens-newswire.com. Retrieved 2012-11-14.

16.6   Further reading

- Winterhalder, B., R. Larsen, and R. B. Thomas. (1974). "Dung as an essential resource in a highland Peruvian community". *Human Ecology* **2** (2): 89–104. doi:10.1007/BF01558115.

16.7   External links

- Application and environmental risks of livestock manure

- North American Manure Expo

- Cornell Manure Program

- County-Level Estimates of Nitrogen and Phosphorus from Animal Manure for the Conterminous United States, 2002 United States Geological Survey

- Manure Management, Water Quality Information Center, U.S. Department of Agriculture

- Livestock and Poultry Environmental Learning Center, an eXtension community of practice about animal manure management

- Antibiotics and Hormones in Animal Manure (Webcast): A two part webcast series about the science available on potential risks and best management practices related to antibiotics and hormones from animal manure

Manure on a wall.

Chapter 17

Manure management

Manure management refers to capture, storage, treatment, and utilization of animal manures in an environmentally sustainable manner. It can be retained in various holding facilities. Animal manure (also referred to as animal waste) can occur in a liquid, slurry, or solid form. It is utilized by distribution on fields in amounts that enrich soils without causing water pollution or unacceptably high levels of nutrient enrichment. Manure management is a component of nutrient management.

17.1 See also

- Manure

- Concentrated Animal Feeding Operation

- Animal feeding operation

- Slurry pit

17.2 References

Report for Congress: Agriculture: A Glossary of Terms, Programs, and Laws, 2005 Edition

17.3 External links

- Animal Manure Management on http://eXtension.org

- Animal Manure Management on http://eXtension.org

- Animal Manure Nutrient Management on http://eXtension.org

Chapter 18

Maxicrop

Maxicrop is a company that sells seaweed-based liquid or powdered organic fertiliser, promoted as a "bio-stimulant".

Doug Edmeades, a New Zealand scientist and agricultural consultant, was sued by Maxicrop when he said that their product was useless.[1]

The resulting case Bell-Booth Group Ltd v Attorney-General reviewed evidence provided by the Ministry of Agriculture established that Maxicrop did not materially improve the growth of plants and supported Dr. Edmeades' claims of inefficacy.

18.1 References

[1] Morgan, Jon (2008-06-06). "Outspoken scientist follows conscience". The Press.

18.2 External links

- Maxicrop International

Chapter 19

Milorganite

Milorganite is the trademark of a biosolids fertilizer produced by the Milwaukee Metropolitan Sewerage District.[1] The District captures wastewater from the Milwaukee metropolitan area, including local industries. This water is then treated at the Jones Island Water Reclamation Facility in Milwaukee, Wisconsin with microbes convert sewage solids into sewage sludge. Cleaned water is then returned to Lake Michigan. The recycled organic nitrogen fertilizer, which is sold throughout North America, reduces the need for manufactured nutrients. After more than 90 years, it is one of the largest and most continuous examples of such programs.[1][2][3][4]

19.1 History

Milorganite's history began with Milwaukee's goal to clean up its rivers and Lake Michigan. Rather than land filling sludge, the byproduct was used in a pioneering effort to make, distribute and sell fertilizer.[4] "Its production is among the largest recycling programs in the world."[3][5]

The Jones Island Plant was among the first sewage treatment plants in the United States to succeed in using the activated sludge treatment process.[6] "It was the first treatment facility to economically dispose of the recovered sludge by producing an organic fertilizer." In the early 1980s the plant needed extensive reworking, "this does not detract from its historic significance as a pioneering facility in the field of pollution control technology."[5] "The world's first large scale wastewater treatment plant was constructed on Jones Island, near the shore of Lake Michigan."[7] It had the largest capacity of any plant in the world when constructed.[8] The 1925 plant has been designated as a Historic Civil Engineering Landmark by the American Society of Civil Engineers.[6][9][10]

The name Milorganite is a concatenation of the phrase *Milwaukee Organic Nitrogen,* and was the result of a 1925 naming contest held in National Fertilizer Magazine. Raising taxes for public health was relatively controversial in the early 1900s. In 1911, reform minded socialists were elected on a platform calling for construction of a wastewater treatment plant to protect against water borne pathogens.[11][12] Experiments showed that heat-dried activated sludge pellets "compared favorably with standard organic materials such as dried blood, tankage, fish scap, and cottonseed meal."[13] Sales to golf courses, turf farms and flower growers began in 1926.[14] Milorganite was popularized during the 1930s and 1940s before inorganic urea became available to homeowners after WWII. With the help of researchers in the College of Agriculture at the University of Wisconsin, the use of waste solids (*i.e.,* activated sludge) as a source of fertilizer was first developed in the early 20th century.[2]

Since its development in 1926 as the first pelletized fertilizer in the United States, Milorganite has sold over 9,000,000,000 pounds (4.1×10^9 kg) of recycled waste. As of 2012, the plant produced about 45,000 tons from heat-dried microbes per year.[3] In combining concerns for the environment and social justice, while successfully navigating the fluctuations and vagaries of the changing waste stream[upper-alpha 1] to deliver an important product through recycling, Milorganite has been at the forefront of the industry, even as it balances conflicting goals.[11]

As the organization itself notes:

"Headquartered in Milwaukee, Wisconsin, Milorganite products are manufactured and marketed by the Milwaukee Metropolitan Sewerage District (MMSD), a regional government agency whose primary focus is providing water reclamation and flood management services for about 1.1 million customers in 28 communities in the Greater Milwaukee Area. Since 1926 MMSD has been a world leader in supplying Organic Nitrogen fertilizers for professional and residential use. While revenue generated through the sale of Milorganite products does not make up for the entire cost to produce and market, our belief in beneficial reuse and recycling makes producing our value added products the clear choice."[1]

The sale of product does not entirely generate sufficient funds to cover the costs of manufacture, but the organization suggests the environmental benefits are a legitimate offsetting consideration.[4]

19.2 Product

Heat-dried biosolids contain slow release organic nitrogen, largely water-insoluble phosphorus bound with iron and aluminum and high organic matter.[15]

Milorganite contains virtually no salts, so it never burns plants – even in the hottest temperatures and driest conditions. It may be applied without water, and is moisture activated at a later time. Each application feeds for 8–10 weeks, resulting in fewer applications.[16]

Milorganite can be used without restriction on gardens intended for human consumption under USEPA rules.[16] There are concerns about contaminants, including waste pharmaceuticals, drug pollution, metals, etc.[17] To alleviate those concerns, Milorganite is tested regularly, more often than is required by governmental regulations. According to the Material safety data sheet it is "registered for sale in all 50 states and meets all federal and state requirements."[16][18]

The United States Department of Agriculture (USDA) certifies milorganite as a biobased product because it is derived from 85% renewable materials.[1] Milorganite is not certified for use on U.S.D.A. organic farms.[17]

University research confirms anecdotal evidence that applying Milorganite on lawns and near plants deters deer due to its odor.[16] However, the manufacturer cannot market Milorganite as a deer repellent because it is not registered as a "pesticide.[upper-alpha 2] The size of the expenditure and the lack of a guaranteed return on investment and its timing was deemed to make the venture not worth it. Therefore, repelling hungry deer from Hostas remains an "off-label" use.[7][17][19][20] Milorganite is tested daily for the presence of heavy metals and pathogens. Milorganite complies with the Environmental Protection Agency's (EPA) "Exceptional Quality" criteria, which establishes the strictest concentration limits in the fertilizer industry for heavy metals, allowing Milorganite to be used on food crops. These limits are based upon extensive risk analysis for a variety of pathways.[21]

19.3 See also

- Water resource management

19.4 References

Notes

[1] For example, the loss of much of the Malting and Brewing industry (*e.g.*, Schlitz Brewing Company and Pabst Brewing Company) that was located in Milwaukee affected both the quality and the quantity of raw material available.[11]

[2] The projected cost of testing to satisfy the requirements of the Environmental Protection Agency was estimated to be between $1 and $2 million, when the entire advertising and public contact budget for the organization is only $2 to $3 million per year. Studies paid for by the water district at the University of Georgia and Cornell showed it to be effective in deterring deer.[19]

Further reading[5]

- Alvord, John W.; Whipple, George C.; Eddy, Harrison P. (April 25, 1911). *A Report to the Common Council upon the Disposal of the Sewage and the Protection of the Water Supply of the City of Milwaukee. Unpublished.* Milwaukee Metropolitan Sewerage District.

- Eddy, Harrison (April 17, 1924). "Sewerage and sewage disposal". *Engineering News-Record* **92** (16). pp. 693–695.

- Gurda, John (May 1978). "Change at the River Mouth: Ethnic. Succession on Milwaukee's Jones Island, 1700 to 1922. (Unpublished master's thesis)". Milwaukee, Wisconsin: University of Wisconsin–Milwaukee.

- Horvath, R. Dennis (May 1964). "The Sewage Disposal Controversy: A Study in Milwaukee Area Politics. (Unpublished master's thesis)". Milwaukee, Wisconsin: University of Wisconsin–Milwaukee.

- Leary, Raymond D.; Peot, Werner A. (1973). "Development of a Wastewater Treatment System for the Milwaukee Metropolitan Sewerage District".

- National Register of Historic Places nomination form for the Jones Island Wastewater Treatment Plant.[5]

Citations

[1] "About us". Milorganite/Milwaukee Metropolitan Sewerage District. Retrieved March 28, 2014.

[2] "History of Milorganite". Milorganite/Milwaukee Metropolitan Sewerage District. Retrieved March 26, 2014.

[3] Tanzilo, Bobby (28 September 2012). "Urban spelunking: Brewing up Milorganite". RSS Feed/OnMilwaukee.com. Retrieved 28 March 2014.

[4] "Milorganite Reaches 9 Billion Pounds with 85 Years of Recycling" (Press release). Milwaukee, Wisconsin: PRWEB. June 2, 2012. Retrieved March 26, 2014.

[5] Merritt, Raymond H. (1982). *Historical Report Photographs, Written Historical and Descriptive Data* (PDF). *Historic American Engineering Record* (National Park Service). Retrieved April 1, 2014.

[6] "Environmental Draft Impact Statement: Milwaukee Metropolitan Sewage District; Water Pollution Abatement Program, No. E1S801072DB". Environmental Protection Agency. November 1980. p. V-100. Retrieved April 1, 2014.

[7] Stephens, Odin L.; Mengak, Michael T.; Osborn, David; Miller, Karl V. (March 2005). "Using Milorganite to temporarily repel white-tailed deer from food plots" (PDF). *Wildlife Management Series* (2). University of Georgia Warnell School of Forestry and Natural Resources. Retrieved April 1, 2014.

[8] Freese, Simon W., P.E.; Sizemore, Deborah Lightfoot. *A Century in the Works: 100 Years of Progress in Civil and Environmental Engineering; Freese and Nichols Consulting Engineers 1894-1994* (PDF). p. 44. Retrieved April 2, 2014.

[9] American Society of Civil Engineers (August 13, 1974). "Regarding designation of the Jones Island plant as a national engineering landmark" (Press release).

[10] Program Management Office, Milwaukee Water Pollution Abatement Program; CH2M HILL, INC.; Donohue & Associates, Inc.; Howard Needles Tammen & Bergendoff; Graef, Anhalt, Schloemer and Associates, Inc.; Poly tech, Inc.; J.C. Zimmerman Engineering Corp.; Camp Dresser & McKee, Inc. (April 1982). *Historic Documentation of the Jones Island West Plant* (PDF). *Milwaukee Water Pollution · Abatement Program* (Milwaukee, Wisconsin: Milwaukee Metropolitan Sewerage District). Retrieved August 8, 2015.

[11] Foote, Stephanie, Ed.; Mazzolini, Elizabeth; Schneider, Daniel (Chapter 7) (2012). "7, "Purification or Profit: Milwaukee and the Contradiction of Sludge". *Histories of the Dustheap: Waste, Material Cultures, Social Justice*. Cambridge, Massachusetts: MIT Press. pp. 171–197. ISBN 0-262-01799-7. Retrieved March 26, 2014.

[12] Mortimer, Clifford (May 1981). "The Lake Michigan Pollution Case: A Review and Commentary on the Limnological and Other Issues". *Publications of the Great Lakes Center for Research* (Milwaukee, Wisconsin: Center for Great Lakes Studies, University of Wisconsin – Milwaukee): 2–3. Retrieved March 29, 2014..

[13] Eleventh Annual Report of the Sewerage Commission of the City of Milwaukee for 1924, pp. 32–42.

[14] See, North American's Most Widely Known, Respected, and Beloved Turfgrass Agronomist, The O.J. Noer Research Foundation, Inc., Michigan State U. Libraries, Turfgrass Information Center, www.lib.msu.edu/tgif.

[15] Miller, Matt; O'Connor, George A. (2009). "Longer-term Phytoavailability of Biosolids-Phosphorus". *Agronomy Journal* (101): 889–896. or puri.fcla.edu/fcia/etd/UFE0022710

[16] Fedigan, Lamont. *21st Century Homestead: Organic Farming*. p. 112. Retrieved August 8, 2015.

[17] Harrison, Ellen Z. Director (2006). "Fact Sheet 2006: Home Garden Use of Milorganite" (PDF). Ithaca, NY: Cornell University Waste Management Institute. Retrieved March 28, 2014.

[18] "Material Safety Data Sheet (MSDS)". Milorganite. Retrieved April 3, 2014.

[19] Behm, Don (January 18, 2009). "EPA derails plans to market Milorganite as deer repellent". *Milwaukee Journal-Sentinel*. Retrieved March 29, 2014.

[20] "Newsletter". National Biosolids Partnership. January 22, 2009. Retrieved March 28, 2014.

[21] "Standards for the Use and Disposal of Sewage Sludge" (PDF). *Region 10: The Pacific Northwest*. United States Environmental Protection Agency.

19.5 External links

- Milorganite Webpage

- History of Milorganite

- Washington State Department of Agriculture Fertilizer and Pesticide Database — "M"

- USDA Bio-Preferred Program

- "Milorganite informational video" (video). Milorganite. March 15, 2012. Retrieved March 29, 2014.

Chapter 20

Olive mill pomace

Olive mill pomace is a by-product from the olive oil mill extraction process. Usually it is used as fuel in a cogeneration system or as organic fertiliser after a composting operation.

Olive mill pomace compost is made by a controlled biologic process that transforms organic waste into a stable humus. Adding composted olive mill pomace as organic fertiliser in olive orchards allows the soil to get nutrients back after each olive crop.

20.1 Two-phase pomace

In crude olive oil production, the traditional system, i.e. pressing, and the three-phase system produce a press cake and a considerable amount of waste water while the two-phase system, which is mainly used in Spain, produces a paste-like waste called "alperujo" or "two-phase pomace" that has a higher water content and is more difficult to treat than traditional solid waste. The water content of the press cake, composed of crude olive cake, pomace and husk, is about 30 percent if it is produced by traditional pressing technology and about 45–50 percent using decanter centrifuges. The press cake still has some oil that is normally recovered in a separate installation. The exhausted olive cake is incinerated or used as a soil conditioner in olive groves.

20.2 External links

- Does the composted olive mill pomace increase the sustainable N use of olive oil cropping?. 2009. 16th Nitrogen workshop. Connecting different scales of Nitrogen use in agriculture. Turin Italia.

Chapter 21

Riverm

Riverm is an ecologically safe, liquid organic fertilizer which was developed in Ukraine by Ukraine's International Ecological Foundation AQUA-VITAE and National Agrarian University of Ukraine.

"Riverm" successfully passed government testing and was registered in Ukraine on 6.06.2005 under registration number 1921 (in certificate series A №01031), and also was approved by an international organization - System of Independent Certification (SIC) as an ecologically clean fertilizer which complies with the international standard of ISO 14024:1999.

In 2006 "Riverm" became the laureate of national distinction «Vishcha proba» in Ukraine.

Since 2007 IEF AQUA-VITAE has been accepted into the inspection-certification program "Organic production" by the international group Control Union Certifications and acquired the right to mark the product as conforming to EEC2092/91 and CUC Inputs standards.

In 2008 "Riverm" became the winner of Ukrainian «Best product of the year» award.

Today "Riverm" is available for sale not only in Ukraine, but also across the world.

21.1   References

- Principles of environmentally safe agriculture - AQUA-VITAE. Риверм

Chapter 22

Phosphate rich organic manure

Phosphate rich organic manure is a type of fertilizer used as an alternative to diammonium phosphate and single super phosphate.

Phosphorus is required by all plants but is limited in soil, creating a problem in agriculture. In many areas phosphorus must be added to soil for the extensive plant growth that is desired for crop production. Phosphorus was first added as a fertilizer in the form of single super phosphate (SSP) in the mid-nineteenth century, following research at Rothamsted Experimental Station in England. SSP is non-nitrogen fertiliser containing Phosphate in the form of monocalcium phosphate and Gypsum which is best suited for Alkali soils to supplement Phosphate and reduce soil alkalinity.[1]

The world consumes around 140 million tons of high grade rock phosphate mineral today, 90% of which goes into the production of diammonium phosphate (DAP). Excess application of chemical fertilizers in fact reduces the agricultural production as chemicals destroy natural soil flora and fauna. When DAP or SSP is applied to the soil only about 30% of the phosphorus is used by the plants, while the rest is converted to forms which cannot be used by the crops [x1,X2], a phenomenon which is known as phosphate problem to soil scientists.

Directly mixing finely ground rock phosphate mineral into organic manure produces a fertilizer known as **phosphate rich organic manure** (PROM). Research indicates that this substance may be a more efficient way of adding phosphorus to soil than applying chemical fertilizers.[2][3] Other benefits of PROM are that it supplies phosphorus to the second crop planted in a treated area as efficiently as the first, and that it can be produced using acidic waste solids recovered from the discharge of biogas plants.

Phosphorus in rock phosphate mineral is mostly in the form of tricalcium phosphate, which is water-insoluble. Phosphorus dissolution in the soil is most favorable at a pH between 5.5 and 7.[4] Ions of aluminum, iron, and manganese prevent phosphorus dissolution by keeping local pH below 5.5, and magnesium and calcium ions prevent the pH from dropping below 7, preventing the release of phosphorus from its stable molecule.[4] Microorganisms produce organic acids, which cause the slow dissolution of phosphorus from rock phosphate dust added to the soil, allowing more phosphorus uptake by the plant roots. Organic manure can prevent ions of other elements from locking phosphorus into insoluble forms. The phosphorus in phosphate enhanced organic manure is water-insoluble, so it does not leach into ground water or enter runoff [x]

Most phosphate rocks can be used for phosphate rich organic manure. It was previously thought that only those rocks which have citric acid soluble phosphate and those of sedimentary origin could be used.[3] Rocks of volcanic origin can be used as long as they are ground to very fine size.

Organic manure should be properly prepared for use in agriculture, reducing the C:N ratio to 30:1 or lower. Alkaline and acidic soils require different ratios of phosphorus.

PROM is known as a green chemistry phosphatic fertilizer. Addition of natural minerals or synthetic oxides in water-insoluble forms that contain micronutrients such as copper, zinc, and cobalt may improve the efficiency of PROM. Using natural sources of nitrogen, such as *Azolla*, may be more environmentally sound.[5]

22.1 Phosphate Rich Organic Manure under FCO

Ministry of Agriculture and Cooperation, Government of India has now approved Phosphate Rich Organic Manure (PROM) and included the same under Fertilizer Control Order (FCO). The approved specifications may be seen from Gazzetee Notification from the web site of PROM Society here: http://www.promsociety.net/

22.2 References

[1] "Single Superphosphate" (PDF). Retrieved 11 March 2014.

[2]

[3]

[4] Brady, NC, Nature and Properties of Soils, Collier Macmillan, London,1984.

[5]

- http://www.sciencedirect.com/science?_ob=ArticleURL&_udi=B6V78-40CJYS8-17&_user=10&_rdoc=1&_fmt= &_orig=search&_sort=d&_docanchor=&view=c&_searchStrId=1048602170&_rerunOrigin=google&_acct=C000050221& _version=1&_urlVersion=0&_

- http://precedings.nature.com/documents/2411/version/1* http://www.saber.ula.ve/bitstream/123456789/29069/1/ mineral.pdf

- http://www.uctm.edu/journal/j2006-3/05-Ivanova-297-302.pdf

- Narayanan, C. M. (2012). "Production of Phosphate-Rich Biofertiliser Using Vermicompost and Anaerobic Digestor Sludge—A Case Study". *Advances in Chemical Engineering and Science* **2** (2): 187–191. doi:10.4236/aces.2012.22022.

- Principles of Phosphate Fertilization and PROM – Progress Review 2012

https://www.researchgate.net/publication/235918492_Principles_of_Phosphate_Fertilization_and_PROM__Progress_ Review_2012?ev=prf_pub

Chapter 23

Rockdust

A coal miner in West Virginia spraying rockdust to reduce the combustible fraction of coal dust in the air of a bituminous coal mine.

Rock dust, also known as **rock powders**, **rock minerals**, **rock flour**, **soil remineralization**, and **mineral fines**, consists of finely crushed rock, processed by natural or mechanical means, containing minerals and trace elements widely used in organic farming practices.

The igneous rocks basalt and granite often contain the highest mineral content, whereas limestone, considered inferior in this consideration, is often deficient in the majority of essential macro-compounds, trace elements, and micronutrients.

Rock dust is not a fertilizer, for it lacks the qualifying levels of nitrogen, potassium, and phosphorus.

Rockdust is also the limestone-based product sprayed on walls inside underground coal mines to keep coal dust levels down. This is to prevent coal dust explosions and also to prevent the incidence of black lung disease.

23.1   Background

Soil remineralization creates fertile soils by returning minerals to the soil which have been lost by erosion, leaching, and or over-farming. It functions the same way that the Earth does: during an Ice Age, glaciers crush rock onto the Earth's soil mantle, and winds blow the dust in the form of loess all over the globe. Volcanoes erupt, spewing forth minerals from deep within the Earth, and rushing rivers form mineral-rich alluvial deposits.

Rock dust is added to soil to improve fertility and has been tested since 1993 at the Sustainable Ecological Earth Regeneration Centre (SEER Centre) in Straloch, near Pitlochry, in Perth and Kinross, Scotland.[1] Further testing has been undertaken by James Cook University, Townsville, Far North Queensland.[2]

23.2   Research

SEER's research claims that the benefits of adding rockdust to soil include increased moisture-holding properties in the soil, improved cation exchange capacity and better soil structure and drainage. Rockdust also provides calcium, iron, magnesium, phosphorus and potassium, plus trace elements and micronutrients. By replacing these leached minerals it is claimed that soil health is increased and that this produces healthier plants.

Typical composition table of rockdust

23.2.1   Available silicon

Silicon is thought to be the major element effecting the strength of cell wall development. However it is the amount of available silica that has a dramatic effect on the plant strength and subsequent health. To highlight this, plants that are grown in very sandy soils, (being high in non available silica), often require a silica based fertiliser to provide available silicon.

Silicon comes in silicon multi-oxide molecules (e.g. SiO_2, SiO_4, SiO_6, and SiO_8). Each molecule shape is thought to pack in different ways to allow different levels of availability.

23.2.2   Phosphate fixation

Often phosphorus is locked in soils due to many years of application of traditional fertilisers. The use of micronutrient-rich fertiliser enables plants to access locked phosphorus.

23.2.3   Paramagnetism

The elements high in available 2+ valence electrons, calcium, iron and magnesium in particular contribute to paramagnetism in soil which aid in cation exchange capacity.

23.2.4   pH

The calcium and magnesium in high quality have the ability to neutralise pH in soils, in effect acting as a liming agent.

23.2.5   Spreading and applying dust in agriculture

Rockdust can be applied to soil by hand application, via broadcast spreader or by fertigation. Where possible the rockdust can be worked into the ground either physically or by using water to wash in.

23.2.6 Rate of application

In some soils which display poor levels of nutrients, application rates of 10 tonnes per hectare are required. In Australia, namely the Riverland, Riverina, Langhorne Creek, Barossa and Mclaren Vale regions, rates are 3–5 tonnes per hectare. In a garden application, this might equate to 400 grams per square metre.

23.3 History

Rockdusting, also known as soil remineralization, was mentioned in the 19th-century book *Bread From Stones* by chemist Julius Hensel.

In the 20th century, rockdusting was popularized by science writers John D. Hamaker, Larry Ephron, Alden Bryant, Don Weaver, Harvey Lisle, Arden Andersen and Lee Klinger.[3] Klinger uses remineralization through rock dust to restore ailing trees.[4]

23.4 See also

- Greensand, used as a source of potassium

- Organic farming

- Organic fertilizer

- Phosphorite, also called rock phosphate

- Rock flour

- Soil conditioner

23.5 References

[1] Remineralization Might Save Us From Global Warming, *The Independent*, Paul Kelbie, 21 March 2005

[2] De Silva, Meragalge Swarna Damayanthi Luxmei. "The effects of soil amendments on selected properties of tea soils and tea plants (Camellia sinensis L.) in Australia and Sri Lanka.". *James Cook University*. James Cook University. Retrieved 25 April 2015.

[3] Klinger, Lee (29 December 2007). "Minerals for Aging Soils". *Remineralize the Earth*. Remineralize.org. Retrieved 11 March 2014.

[4] Rich, Deborah (29 October 2005). "OAK LORE". *SF GATE*. San Francisco Chronicle. Retrieved 11 March 2014.

Chapter 24

Seaweed fertiliser

Seaweed fertiliser, also spelled **seaweed fertilizer**, is seaweed that is collected and used as fertilizer.

In the Channel Islands, such seaweed fertiliser is known as **vraic** in their dialects of Norman, a word that has also entered Channel Island English, the activity of collecting vraic being termed *vraicking*. In Scotland, it is used as fertiliser in lazybeds or *feannagan*. Falkland Islanders have also been nicknamed "Kelpers" from time to time, from collecting seaweed partly for this purpose.[1]

Several of the 12,000+ varieties of seaweed in the ocean have been shown to be valuable additions to the organic garden and can be abundantly available free for those living near the coast. However, caution should be observed when collecting seaweed, particularly from areas that are liable to pollution, such as downriver (including estuaries) of industrial activities as seaweed is susceptible to contamination. There are also legal implications relating to gathering seaweed, and concerns about sustainability[2]

A perhaps less serious potential problem with seaweed is its salt content. While it is unlikely to add sufficient seaweed to seriously upset the balances of salt in the soil, it is not liked by worms, who will not live in it. It can be hosed down before adding to the soil to reduce the salt content, or left to be desalinated by rainwater. Rinsing seaweed is risky as valuable alginates are potentially lost to runoff.

Seaweed, particularly bladderwrack, kelp or laminaria, can be either applied to the soil as a mulch (although it will tend to break down very quickly) or can be added to the compost heap, where it is an excellent activator.[3] In terms of soil structure it does not add a great deal of bulk, but its jelly like alginate content helps to bind soil crumbs together, and it contains all soil nutrients (0.3% N, 0.1% P, 1.0% K, plus a full range of trace elements) and amino acids. For those who cannot gather fresh seaweed, it is available commercially in a dried 'meal' form or as a concentrated liquid extract which is active in significantly smaller rates. While 'meal' products are limited to soil applications due to their insolubility, foliar feeding or root zone applications through drip lines are possible with soluble extracts.

24.1 See also

- Organic fertilizer
- Organic gardening
- Organic farming
- Rockdust

24.2 References

[1] "The people of the Falklands". Stanford University. Retrieved 2009-05-19.

Vraicking in Jersey

[2] "Seaweed products". *Gardening Advice*. Royal Horticultural Society. 2009-03-31. Retrieved 2009-05-19.

Former feannagan *on Great Bernera, Outer Hebrides, which would have been improved with seaweed fertiliser*

[3] Coleby-Williams, Jerry (2006-12-16). "Fact Sheet: Seaweed Fertiliser". *Gardening Australia*. ABC. Retrieved 2009-05-19.

Chapter 25

Slurry pit

A **slurry pit**, also known as a farm slurry pit, **slurry tank**, **slurry lagoon** or **slurry store**, is a hole, dam, or circular concrete structure where farmers gather all their animal waste together with other unusable organic matter, such as hay and water run off from washing down dairies, stables, and barns, in order to convert it, over a lengthy period of time, into fertilizer that can eventually be reused on their lands to fertilize crops.[1][2] The decomposition of this waste material produces deadly gases, making slurry pits potentially lethal where precautions (separate air supply etc.) are not taken.[3]

25.1 Nutrient value

The liquid manure blend, or slurry, can be a rich source of nitrogen, phosphorus, and potassium.

25.2 Risks

Slurry pits present multiple risks. Risk of drowning from falling into the tank, or risk from the gases produced during the decomposition. The gases include methane, carbon dioxide, ammonia and hydrogen sulphide, these are heavier than air, so in a pit will not disperse quickly. Some are odourless or can destroy sense of smell (thus becoming odourless). If inhaled, they can lead to rapid unconsciousness (which could then lead to drowning), are poisonous and can suffocate.[4][5][6][7][8][9] The health and safety executive of Northern Ireland specifies working in a slurry pit as specialist work, requiring separate air supply and a line connecting the worker to two additional people outside of the tank.[3]

25.3 See also

- Manure management

25.4 References

[1] "What is a slurry tank? | Dairy Farming Facts". Thisisdairyfarming.com. Retrieved 2013-06-24.

[2] "Slurry lagoons | Dairy Farming Facts". Thisisdairyfarming.com. Retrieved 2014-06-08.

[3] "Slurry Gases Can Kill" (PDF). HSENI.

[4] "The lethal risks of working with slurry". BBC News Online. 17 September 2012.

[5] Boy drowned in Shropshire slurry pit, inquest told BBC News Online 2010-06-09

[6] Two men killed in Essex farm slurry pit **BBC News Online** 2011-07-16

[7] Slurry tank tragedy: Rugby star Nevin Spence killed **BBC News Online** 2012-09-16

[8] "Boy dies in slurry tank accident near Dunloy, County Antrim". **BBC News Online**. 8 June 2014.

[9] "Dunloy farm death: What is slurry and why is it so dangerous?". **BBC News Online**. 8 June 2014.

25.5 Text and image sources, contributors, and licenses

25.5.1 Text

- **Organic fertilizer** *Source:* https://en.wikipedia.org/wiki/Organic_fertilizer?oldid=677264925 *Contributors:* Julesd, Robbot, Tobias Bergemann, Discospinster, Rich Farmbrough, Alansohn, Wtmitchell, Velella, Pauli133, Tobyc75, Cskin, Bobrayner, DVdm, Triku~enwiki, Wavelength, Haranoh, CWenger, SmackBot, C.Fred, Stepa, Gilliam, Chris the speller, Bidgee, Jefffire, Smokefoot, Drphilharmonic, Minna Sora no Shita, Phatom87, Dawnseeker2000, MER-C, Doctorhawkes, Ja 62, Inwind, Moogwrench, Red58bill, Flyer22, ClueBot, SuperHamster, Alandmanson, BOTarate, Katalyst2, XLinkBot, Rror, Addbot, MrOllie, Download, Tide rolls, Luckas-bot, Yobot, Synchronism, AnomieBOT, Lcemrich, CeciliaPang, Disagreeableneutrino, Xqbot, Apothecia, Anna Frodesiak, Doulos Christos, 1243trel;ghsfdk;lgsfd, Exploredotorg, HamburgerRadio, Pinethicket, Rushbugled13, Codwiki, MastiBot, Wien12, Mean as custard, Ripchip Bot, Slon02, Cupidophone, Look2See1, Iwan Novirion, Sasuke43, AManWithNoPlan, Δ, Donner60, Cobaltriposte, 28bot, ClueBot NG, Auchansa, Widr, Mam1975, Ramaksoud2000, Mnbappy25, BattyBot, Organicgg, Greenbean2006, Jaysakle, Sidelight12, Frosty, AcquiesceLife, Moms dragons, SamanthaPuckettIndo, Vastaradelia, Dsprc, Trackteur, Seekfertilizer, EvMsmile, Krishibid roy and Anonymous: 94

- **Azomite** *Source:* https://en.wikipedia.org/wiki/Azomite?oldid=671717598 *Contributors:* DragonflySixtyseven, Vsmith, Ground Zero, Deli nk, Huon, WhatamIdoing, Legobot, AnomieBOT, Citation bot, Maitchy, Benzband, BattyBot, Khazar2, Victoriasays, Jackmcbarn, Monkbot and Anonymous: 3

- **Bioeffector** *Source:* https://en.wikipedia.org/wiki/Bioeffector?oldid=628298945 *Contributors:* DragonflySixtyseven, Wavelength, Raupp, WOSlinker, Ironholds, FrescoBot and EmausBot

- **Biofertilizer** *Source:* https://en.wikipedia.org/wiki/Biofertilizer?oldid=678913108 *Contributors:* Bearcat, Rpyle731, Klemen Kocjancic, Nekochan, Richwales, Bgwhite, Yamaguchi⬜⬜, Frankahilario, Teratornis, Wikid77, Raupp, Dougher, Lenticel, R'n'B, Lerdthenerd, SieBot, Harry~enwiki, EoGuy, Porchcorpter, Roxy the dog, Addbot, Quarzyr, Sugeesh, Luckas-bot, Yobot, AnomieBOT, Jim1138, Xqbot, INeverCry, D'ohBot, Avanti j, Bgpaulus, Rweathers, EmausBot, GoingBatty, ZéroBot, GreenPhosphateBacteria, Greenphosphatemicrob, GREENPHOSPHATEmicrobes, ClueBot NG, Abrahamathistan, CantabrigianAlicia, Joydeep, Pabos95, Sidelight12, Chithra Bivin, Amruta Bartakke, Matthewjparker, Learnerktm, Kathyab1 and Anonymous: 35

- **Biosolids** *Source:* https://en.wikipedia.org/wiki/Biosolids?oldid=681280694 *Contributors:* Stone, Ike9898, BesigedB, Christopherlin, Spiffy sperry, CanisRufus, Reinyday, Cmdrjameson, Vortexrealm, Alansohn, Paleorthid, Woohookitty, Tabletop, Rjwilmsi, Trlovejoy, Ground Zero, Jrtayloriv, Stevenfruitsmaak, The Rambling Man, Wavelength, RussBot, Rhallanger, NeilN, SmackBot, Verne Equinox, Deli nk, Gobonobo, Nehrams2020, Dekr, Sanjour, Moreschi, Bheal, Gregalton, MaxPont, DraiconeBot, Justanother, Nono64, Ncmvocalist, Paulbracegirdle, Ned Beecher, John Carter, Bcrathorne, Enviroboy, Red58bill, Ladidie, Seismas, Pakaraki, Mild Bill Hiccup, Thewellman, Thingg, Dthomsen8, Fhearl, Ericsthornton, SpBot, Lightbot, Worksafe, Plasticbot, AnomieBOT, SamuelOBruhl, FrescoBot, Kwiki, Wikibiogas, Kayjenney, Jim Bynum, SoledadKabocha, Prymshbmg, EvMsmile, Gruster, CndTroubadour and Anonymous: 27

- **Chicken manure** *Source:* https://en.wikipedia.org/wiki/Chicken_manure?oldid=680906222 *Contributors:* Dennis Brown, Nikkimaria, Deli nk, IronGargoyle, LadyofShalott, NE Ent, MelanieN, Red58bill, SimonTrew, Xqbot, ChildofMidnight, Trappist the monk, Zanhe, ClueBot NG, This lousy T-shirt, Northamerica1000 and Anonymous: 2

- **Compost** *Source:* https://en.wikipedia.org/wiki/Compost?oldid=682692287 *Contributors:* Eloquence, Bryan Derksen, -- April, Andre Engels, Jpsturm, Josh Grosse, Anthere, Zoe, Heron, Quercusrobur, Edward, Pit~enwiki, Menchi, Qaz, Paul A, Egil, DavidWBrooks, Stan Shebs, Ronz, Kokamomi, 5ko, Julesd, Glenn, Llull, º¡º, Raven in Orbit, Jengod, Dcoetzee, Ike9898, Jose Ramos, Samsara, Geraki, Mignon~enwiki, Pollinator, UninvitedCompany, Robbot, Moncrief, Goethean, Thunderbolt16, KellyCoinGuy, Antonin~enwiki, Tsavage, Pengo, Alan Liefting, MPF, Elf, Timpo, Bradeos Graphon, Guanaco, Solipsist, Bobblewik, LennartBolks~enwiki, JRR Trollkien, Chowbok, Beland, Madmagic, OverlordQ, Zantolak, Jeremykemp, Ukexpat, Sonett72, Stevenmattern, Discospinster, Vsmith, LindsayH, ESkog, Pedant, Tompw, MisterSheik, *drew, Kwamikagami, EurekaLott, CDN99, Bobo192, Spalding, Fir0002, Duk, Vortexrealm, Cohesion, Giraffedata, Kjkolb, Jeodesic, MPerel, Batneil, Frank101, Alansohn, Anthony Appleyard, Arthena, Paleorthid, Keflavich, Rohirok, Bart133, Velella, Mikeo, Kusma, Gene Nygaard, Dismas, Brookie, Abanima, Velho, Lincspoacher, Doctor Boogaloo, Uncle G, WadeSimMiser, Jwanders, Bennetto, Mandarax, Lego872, Kissekatt, Rjwilmsi, Valentinejoesmith, Harro5, HappyCamper, FlaBot, Eubot, AdnanSa, Latka, Margosbot~enwiki, Angstrom, EronMain, Benanhalt, Ahpook, YurikBot, Wavelength, RobotE, Cheesewire, Sceptre, Kollision, Carllindstrom, Splash, SpuriousQ, Stephenb, GeeJo, Leighblackall, NawlinWiki, Grandad, Nirvana2013, R'son-W, Brandon, Diotti, My Cat inn, Semperf, Natkeeran, DeadEyeArrow, Supspirit, CLW, Wknight94, Closedmouth, Xaxafrad, ArielGold, Whouk, Mejor Los Indios, ChemGardener, Erik Sandberg, SmackBot, Evilsofcompost, Rex the first, Chairman S., WookieInHeat, Edgar181, Cazort, Jonobennett, Gilliam, Donama, Ghosts&empties, Mycota, Andy M. Wang, Chris the speller, Bluebot, Persian Poet Gal, Salvor, Deli nk, ArcaneMachine, Brimba, Rrburke, Nahum Reduta, Nakon, 4hodmt, Drphilharmonic, DMacks, Torst, Byelf2007, Rory096, AThing, Srikeit, Euchiasmus, Gobonobo, Tim Q. Wells, Mr. Lefty, IronGargoyle, FrostyBytes, Werdan7, Metao, Peter Horn, ASVP, KJS77, Hu12, Informedbanker, BranStark, Natronomonas, Iridescent, Russcohn, Tony Fox, Dp462090, Courcelles, Ziusudra, Poolkris, JForget, Sakowski, Makeemlighter, DeLarge, Harej bot, LLucas, DoranM, Funnyfarmofdoom, Fnlayson, Peripitus, Yan24, Lugnuts, Hibou8, Tawkerbot4, Teratornis, Ameliorate!, Gimmetrow, Satori Son, Epbr123, Nahaboy, Marek69, Zé da Silva, Grayshi, Orfen, Dajagr, Derzsi Elekes Andor, Panphage, MadeHere, Jj137, TimVickers, Paul Christensen, Smartse, Gundam07th, Danger, Kent Witham, ScottM84, Kariteh, .alyn.post., JAnDbot, Deflective, Husond, Od1n, DuncanHill, Hut 8.5, TheEditrix2, Magioladitis, Bongwarrior, VoABot II, Yyyikes, CTF83!, Sustainableyes, MartinBot, Arjun01, CommonsDelinker, Artaxiad, J.delanoy, Pharaoh of the Wizards, Rlsheehan, Bogey97, Eliz81, Naniwako, Vanished user g454XxNpUVWvxzlr, DadaNeem, Ontarioboy, Brendan19, FuegoFish, Fiona C Mackenzie, Sjforman, DASonnenfeld, Idioma-bot, Signalhead, Sporti, Jeff G., W. B. Wilson, Barneca, JBazuzi, Philip Trueman, TXiKiBoT, AgamemnonZ, Joe2832, ElinorD, Aymatth2, Corvus cornix, LeaveSleaves, BotKung, Anarchangel, Scabether, Temporaluser, Gnowk, Red58bill, MrChupon, ZBrannigan, Demize, Sauronjim, L32007, SieBot, YonaBot, Yintan, Tttools123, Keilana, Jojalozzo, Robotchampion, OsamaBinLogin, Macnabr, Wmpearl, Antonio Lopez, Emo joe, Bingbing154, Rdoiron1, Hobartimus, Nancy, Dillard421, MitchCallison, Gardenmandy, Macdaddyfolife, Nn123645, Denisarona, ClueBot, GorillaWarfare, Dobermanji, The Thing That Should Not Be, Wysprgr2005, LisaSmegal, Connect 321, Uncle Milty, CounterVandalismBot, Auntof6, Tadhussey, Jcjenkins, OrBot, Excirial, Mindcry, Grey Matter, Fradol, Micha, NJGW, Schwellungswasser, MasterOfHisOwnDomain, Methanus, Ortos12, XLinkBot, Gnowor, Lumenos,

Avoided, Gazimoff, Eroeben, Dkp205w, Apmorrison1993, Jojhutton, Zellfaze, MrOllie, CarsracBot, Bassbonerocks, FCSundae, Arben1951, Splodgeness, Cammos, Loser4132, 102orion, Tide rolls, Totorotroll, Jarble, Martin Hanson, Legobot, Yobot, עידן ד, 2D, Fraggle81, Fm-rauch, THEN WHO WAS PHONE?, Bmwjackson, Dmarquard, BruceMcAdam, AnomieBOT, DemocraticLuntz, 1exec1, Leangreenhome, Jim1138, Golb12, Jan Complace, Goodtimber, Materialscientist, Efefher, Puddypie007, Neurolysis, H9e3k80, Acuares, Apothecia, Wperdue, Jacquibeeadams, Twirligig, Adus123, Memepope313, RibotBOT, Mattg82, Rickproser, Wpnoone, Shadowjams, Noclock, FrescoBot, Lothar von Richthofen, Recognizance, Sickymicky, Winterst, Pinethicket, I dream of horses, Rule 56, Tyman510, Jschnur, Jorden12, FoxBot, Sintau.tayua, Siltloam, Lotje, Wien12, Japsmp, MrX, AbeColey, DARTH SIDIOUS 2, RjwilmsiBot, Tomlauter, DASHBot, John of Reading, Look2See1, Dewritech, RenamedUser01302013, Tommy2010, Winner 42, Mmeijeri, K6ka, Hollylogan, Azecha, Bamyers99, Dr Black Knife, Benjaminoakes, Jscoop5, FloGreen, Mlemacio, Inka 888, BioPupil, Hyronimus299, Agungsuko, VictorianMutant, CharlieEchoTango, Allisonlhjack, Anita5192, Petrb, Will Beback Auto, ClueBot NG, Peter James, Silly popper dude, Gilderien, Stephen.upton, Chester Markel, Mukitil, Tideflat, Cchanak, Nikkijean, Widr, Antiqueight, Avidgardener711, Ckgurney, Helpful Pixie Bot, Micklan, Gob Lofa, Gauravjuvekar, Bofum, BG19bot, Xamnidar, LioRelations, TopDog5450, Northamerica1000, Ealison, Fahdaftab, Badon, Erickthecompostguy, Corteli9, Volcomkewl, Snow Blizzard, BokashiGirl, ChrisBalch, Rudork, Cclehnen, Whitehousee, Palmfarm, EuroCarGT, Whynot777, Yobwej666, OscarK878, Vamfun, Teighbouy, Lugia2453, Frosty, Graphium, Jochen Burghardt, TheRealWallyrus, Corn cheese, Stevethor, TheBigSnax, I am One of Many, Howicus, Blogmaster121, Federales, Ugog Nizdast, Manul, FinallyAUsernameICanUse, Wafflesandcake, Actinomyces, Lagoset, Monkbot, Wolf.KT, Madhattertea, Justin Schaeffer, Madhav Jolly, Thaliaelias, GLG GLG, EvMsmile, KyleCMSmith, Capt.lingard, JMWt, KasparBot, Rm1911, Martaford, Degglenause and Anonymous: 570

- **Cottonseed meal** *Source:* https://en.wikipedia.org/wiki/Cottonseed_meal?oldid=643543516 *Contributors:* Alan Liefting, H-2-O, Creidieki, NebY, SmackBot, Erik9bot, Wikidocudroso, Glorioussandwich, Look2See1, DandelionSteph, Aliabdiali, Monkbot and Anonymous: 5

- **Effluent spreading** *Source:* https://en.wikipedia.org/wiki/Effluent_spreading?oldid=411145682 *Contributors:* Bearcat, Tagishsimon, Canley, Ariold, FrescoBot and Anonymous: 1

- **Feather meal** *Source:* https://en.wikipedia.org/wiki/Feather_meal?oldid=672288923 *Contributors:* Ike9898, Woohookitty, Tabletop, Dialectric, Chris the speller, A5b, Bwpach, Alaibot, Fayenatic london, Fabrictramp, Transisto, Philip Trueman, Zondi, Canis Lupus, Sun Creator, MystBot, Addbot, AnomieBOT, Xqbot, Pinethicket, EmausBot, Akbrown74, PigeonIP, Zenibus and Anonymous: 6

- **Fish emulsion** *Source:* https://en.wikipedia.org/wiki/Fish_emulsion?oldid=531697374 *Contributors:* Ike9898, Toytoy, Vsmith, A2Kafir, Jonnabuz, Josh Parris, Wavelength, Gaius Cornelius, Epipelagic, SmackBot, Uthbrian, Funandtrvl, Capewellmj, DeltaQuad, Pinethicket, Tofutwitch11, Deadlyops, Look2See1, Tommy2010, Jodoax and Anonymous: 7

- **Fish hydrolysate** *Source:* https://en.wikipedia.org/wiki/Fish_hydrolysate?oldid=563970725 *Contributors:* Alan Liefting, BD2412, Epipelagic, SmackBot, Chris the speller, SilkTork, Epbr123, Fratrep, Blueelectricstorm, Yobot, Themfromspace, Look2See1, Mmeijeri, Roger.johan, My name is not dave, MarkRott1 and Anonymous: 4

- **Fish meal** *Source:* https://en.wikipedia.org/wiki/Fish_meal?oldid=684302888 *Contributors:* Ike9898, Dale Arnett, Alan Liefting, O'Dea, Bobo192, MPerel, NTK, Fred J, Rjwilmsi, YurikBot, Epipelagic, SmackBot, Reedy, Edgar181, Jprg1966, Noles1984, CmdrObot, Wegge-Bot, Mirka~enwiki, Cydebot, Centrepull, Jayron32, Nipisiquit, Albany NY, Nyttend, Fleebo, Rafadev, TXiKiBoT, Hqb, Lamro, Legion fi, ClueBot, PixelBot, BOTarate, Jytdog, Dthomsen8, Addbot, Blueelectricstorm, Glane23, CarTick, AnomieBOT, LilHelpa, Obersachsebot, Shadowjams, John of Reading, Look2See1, Sameer.sheik, ZéroBot, Wikiproject1400, EdoBot, Njet002, Whoopdeeda, Jejaramillo, BG19bot, Declangi, Cerabot~enwiki, Tcpretzel, Lizia7, Joseph2302, Tzjones, KasparBot and Anonymous: 35

- **Gomutra** *Source:* https://en.wikipedia.org/wiki/Gomutra?oldid=679877843 *Contributors:* BD2412, Kenfyre, Swaarnim, Widr, Jerodlycett and Anonymous: 2

- **Liquid manure** *Source:* https://en.wikipedia.org/wiki/Liquid_manure?oldid=683555349 *Contributors:* Dialectric, Dravecky, DerBorg, Yobot, AnomieBOT, Northamerica1000, Kevin12xd, 069952497a and JonathanHopeThisIsUnique

- **Manure** *Source:* https://en.wikipedia.org/wiki/Manure?oldid=683621191 *Contributors:* Andre Engels, Karen Johnson, Edward, Julesd, Schneelocke, CAkira, Tpbradbury, SEWilco, Eugene van der Pijll, Robbot, Altenmann, Babbage, Sunray, Hadal, MPF, Everyking, Curps, Wikibob, Sundar, Solipsist, Elmindreda, Kandar, Gadfium, Fuzzy Logic, Antandrus, Onco p53, J3ff, PDH, Mrtrey99, Neutrality, Joyous!, Didactohedron, Dryazan, CALR, DanielCD, Bornintheguz, Rupertslander, Brian0918, CanisRufus, Cybe, Shrike, Pinzo, Chairboy, Dennis Brown, Bobo192, Deathawk, Kghose, Vortexrealm, Krellis, Pharos, Alansohn, Anthony Appleyard, Eric Kvaalen, Arthena, Linmhall, Ninio, Kurt Shaped Box, RoySmith, Bart133, Velella, RainbowOfLight, Sumergocognito, Versageek, Gene Nygaard, Redvers, OwenX, Zelse81, Pixeltoo, Kmg90, Damicatz, JRHorse, MKleid, Dysepsion, Magister Mathematicae, Bruce1ee, Ligulem, Yamamoto Ichiro, Musical Linguist, Jrtayloriv, King of Hearts, Dj Capricorn, WriterHound, Gwernol, Wavelength, Sceptre, Huw Powell, RussBot, Eyeon, Hydrargyrum, Stephenb, Nfu-peng, Rsrikanth05, NawlinWiki, Dialectric, Marknesbitt, Mkill, Jpeob, Ms2ger, Sandstein, Zzuuzz, Chriswaterguy, Allens, NeilN, Paul Erik, Fastifex, CIreland, Tom Morris, SmackBot, Amit A., KnowledgeOfSelf, FloNight, McGeddon, Pgk, Edgar181, Yamaguchi⁇, Hmains, Ghosts&empties, KaragouniS, Persian Poet Gal, Cabazon, Liamdaly620, DHN-bot~enwiki, Can't sleep, clown will eat me, OrphanBot, Onorem, VMS Mosaic, Addshore, Todd unt, Dreadstar, Drphilharmonic, Ligulembot, WhoKnows, Molerat, Kuru, Jidanni, Mgiganteus1, IronGargoyle, Meco, Ryulong, Zapvet, Jose77, RekishiEJ, LadyofShalott, Courcelles, Tawkerbot2, MightyWarrior, J Milburn, Edmont, Sakurambo, Postmodern Beatnik, Makeemlighter, Agemegos, Biscuit xor crunch, Chmee2, Shandris, Montanabw, Abdullahazzam, Gogo Dodo, Chasingsol, Adolphus79, Shirulashem, DumbBOT, Thijs!bot, ManN, Ucanlookitup, Weasel5i2, Dfrg.msc, Uruiamme, AntiVandalBot, Luna Santin, Waterthedog, Zidane tribal, JAnDbot, Tohru Honda13, Fetchcomms, Xewrch, Jheiv, Scholariusx, Pedro, VoABot II, Lucyin, CTF83!, Avicennasis, Giggy, Beagel, Edward321, Esanchez7587, MartinBot, GTZ-44-ecosan, Rob Lindsey, AlphaEta, J.delanoy, Bogey97, Lt. penguin, Maproom, Katalaveno, AntiSpamBot, Myrin1, Toon05, RB972, Treisijs, Guyzero, Inwind, Richard New Forest, Idioma-bot, Birdmaster300, Littleolive oil, Deor, VolkovBot, Leebo, Soliloquial, Philip Trueman, Abtinb, ElinorD, DoversCHAMP, Don4of4, Jackfork, Rakonas, Catzeleven, BabyG14-x, Insanity Incarnate, Bluedenim, PGWG, NHRHS2010, The Random Editor, SieBot, Samsyee, Grimitar, LOOOOSER, Hiddenfromview, Allmightyduck, Mongoviuspa, Antonio Lopez, Faradayplank, Lightmouse, Sgagnon, Denisarona, ClueBot, Ndenison, Drmies, Uncle Milty, DanielDeibler, CounterVandalismBot, Cirt, Puchiko, Excirial, Eeekster, ParisianBlade, Peter.C, Iohannes Animosus, Muro Bot, Sakrechevsky784, DerBorg, Scalhotrod, DumZiBoT, Stickee, Wikiuser100, Cow master5000, Skarebo, Footbalfreek, Jacoemaritz, Willking1979, DOI bot, Neodop, WiiNie, Cuaxdon, Doctormanhattan, Kui97, Glane23, Bassbonerocks, Koppas, Bob K31416, Craigsjones, Tide rolls, Teles, SasiSasi, Jarble, Shwoof94, Otrfan, Loza tarantula tamer, Luckas-bot, Yobot, Librsh, Ploppies, Eric-Wester,

AnomieBOT, Noq, Jim1138, Piano non troppo, Simonwilms3, Materialscientist, CheeseBaron, The High Fin Sperm Whale, Gigemag76, Anna Frodesiak, Almabot, C+C, Tianyamm2, Roxybearskt, Amaury, Martydevall, Smallman12q, Žiedas, Thehelpfulbot, Griffinofwales, Blazeyasha, Ellie Williams, Citation bot 1, Pinethicket, RandomStringOfCharacters, Fumitol, Bogelund, Mr.Yumtheobeselardus, Missylisa153, Some Color Mage, Iheartalesana, EdithBJones, Palangio, Nascar1996, Tbhotch, Whisky drinker, TjBot, Dlarge1125, Connerankerstein, EmausBot, Racerx11, RenamedUser01302013, Jeremyloveshistory, 123901ab, Wikipelli, White Trillium, AceOfClovers2, Tylerman1221, Starwarsfan0066, CowmaNN, Ljohnson13, Ocaasi, L Kensington, MALLUS, Puffin, Carmichael, DASHBotAV, Shitmoo, خالقیان, ClueBot NG, Westonlongjohn, Markagrossman, TruPepitoM, Delusion23, Yakobmetal, CopperSquare, Gaz2k11, Widr, Strike Eagle, Gatorkiter, Gob Lofa, JackieMoon92, Scott98, Juro2351, Georgina007, Northamerica1000, Stidger, Jackmichaelld, Poopoopant, Coalsun, Mark Arsten, Atomician, Snakeslither, Logiefogie1, Gerry Marshall2, BattyBot, Cyberbot II, EuroCarGT, MattSucci, Getfitbiach, Backendgaming, Gbabycakes, AgentHSmith, The Halblood, Trackteur, Liance, EvMsmile, Zombiehunger, KasparBot, Vanessa 122333 and Anonymous: 348

- **Manure management** *Source:* https://en.wikipedia.org/wiki/Manure_management?oldid=525862921 *Contributors:* Altenmann, Trevor MacInnis, Art LaPella, Mion, Agradman, Qwfp, Yobot, Chenry1unl and Anonymous: 1

- **Maxicrop** *Source:* https://en.wikipedia.org/wiki/Maxicrop?oldid=655780841 *Contributors:* Alan Liefting, Dialectric, Mr. Granger, Addbot, AnomieBOT, Look2See1 and Anonymous: 1

- **Milorganite** *Source:* https://en.wikipedia.org/wiki/Milorganite?oldid=681989148 *Contributors:* Ubiquity, Chowbok, Carptrash, Kjkolb, Slightlyslack, Paleorthid, Velella, Gantry, Ground Zero, Alphachimp, McGinnis, Megapixie, Sulfur, Chris the speller, RFD, Mistress Selina Kyle, MrDarwin, Makyen, Walton One, CmdrObot, Ebyabe, Public Menace, Andymarek, Ttgerman, Marcher01, Red58bill, Xnatedawgx, Brookfield53045, 7&6=thirteen, Dthomsen8, Tassedethe, Lightbot, Mfduffy, AnomieBOT, LilHelpa, Enat66, Notanecdotal, Anupmehra, BattyBot, Hamster32, Mirobes, Apacele and Anonymous: 26

- **Olive mill pomace** *Source:* https://en.wikipedia.org/wiki/Olive_mill_pomace?oldid=649633941 *Contributors:* KillerChihuahua, RHaworth, Pdcook, Zefr, Voxii, Japsmp, Look2See1, Julietdeltalima and Anonymous: 1

- **Riverm** *Source:* https://en.wikipedia.org/wiki/Riverm?oldid=546153553 *Contributors:* Alan Liefting, Sanya3, MystBot, Addbot, AnomieBOT and WikitanvirBot

- **Phosphate rich organic manure** *Source:* https://en.wikipedia.org/wiki/Phosphate_rich_organic_manure?oldid=664989622 *Contributors:* Orangemike, Rjwilmsi, IceCreamAntisocial, SmackBot, Chris the speller, CardinalDan, DMR Sekhar, FRSC Chemist, BG19bot and Anonymous: 18

- **Rockdust** *Source:* https://en.wikipedia.org/wiki/Rockdust?oldid=668224634 *Contributors:* Vsmith, Bender235, Mairi, Vortexrealm, Paleorthid, Tabletop, Anomalocaris, Retired username, Mais oui!, SmackBot, Edgar181, Hmains, Lambiam, John, Dl2000, Nydas, MarshBot, Gdexter, Jamesdexter, Nadiatalent, Inwind, Anguswebmaster, Clinchfield, Mild Bill Hiccup, Addbot, Yobot, Boleyn2, Lithopsian, Fisherscreekrockdust, HRoestBot, Zanze123, WikitanvirBot, Helpful Pixie Bot, BG19bot, Northamerica1000, Zollo9999, Dipankan001, USREM, JZNIOSH, Victoriasays, GENAMH, Zombiepaw, Checkmeleon, Magnetar44, SJ Defender and Anonymous: 7

- **Seaweed fertiliser** *Source:* https://en.wikipedia.org/wiki/Seaweed_fertiliser?oldid=645799875 *Contributors:* Tarquin, Quercusrobur, Aarchiba, Alan Liefting, Andycjp, Sam Hocevar, Vortexrealm, Man vyi, RJFJR, Stemonitis, MacRusgail, Gaius Cornelius, Carabinieri, SmackBot, A. B., JackLumber, Seaphoto, Pedro, Wikidoug3, Romfordian, CounterVandalismBot, Niceguyedc, Capewellmj, Winston365, Legobot, AngeloSt9, EmausBot, Look2See1, Matthewcgirling, AvicAWB, Morgan Riley, Theopolisme, Zadradr, NotWith, Mohamed-Ahmed-FG, Pietro13 and Anonymous: 18

- **Slurry pit** *Source:* https://en.wikipedia.org/wiki/Slurry_pit?oldid=628497291 *Contributors:* Tpbradbury, Grunners, Pburka, Ccgrimm, Ninly, Dawd, Jim Derby, Enquire, Dewelar, ImageRemovalBot, Qwfp, Vlambie, Σ, NortyNort, BG19bot, BattyBot, Mogism and Anonymous: 3

25.5.2 Images

- **File:A_manure_storage_silo_in_the_fields_near_Smilde,_Netherlands,_spring_2012.jpg** *Source:* https://upload.wikimedia.org/wikipedia/commons/7/76/A_manure_storage_silo_in_the_fields_near_Smilde%2C_Netherlands%2C_spring_2012.jpg *License:* CC BY-SA 3.0 *Contributors:* Own work *Original artist:* FotoDutch

- **File:Ambox_important.svg** *Source:* https://upload.wikimedia.org/wikipedia/commons/b/b4/Ambox_important.svg *License:* Public domain *Contributors:* Own work, based off of Image:Ambox scales.svg *Original artist:* Dsmurat (talk · contribs)

- **File:Azomite_mineral_ore.jpg** *Source:* https://upload.wikimedia.org/wikipedia/commons/3/3b/Azomite_mineral_ore.jpg *License:* CC BY-SA 3.0 *Contributors:* Photo taken while on tour of the Azomite mineral reserves. https://www.facebook.com/photo.php?fbid=504838392860915&set=a.133289596682465.22370.121643231180435&type=3&theater *Original artist:* Mila Radulovic

- **File:Bedrijfsafval.jpg** *Source:* https://upload.wikimedia.org/wikipedia/commons/3/3a/Bedrijfsafval.jpg *License:* Public domain *Contributors:* ? *Original artist:* ?

- **File:Biosecure_KOI_breeding_and_growing_intensive_facility_in_Israel.jpg** *Source:* https://upload.wikimedia.org/wikipedia/commons/e/e6/Biosecure_KOI_breeding_and_growing_intensive_facility_in_Israel.jpg *License:* CC BY 2.5 *Contributors:* Transferred from en.wikipedia; *Original artist:* Original uploader was Daben2000 at en.wikipedia

- **File:Biosolid.pumpkin.row.jpg** *Source:* https://upload.wikimedia.org/wikipedia/commons/8/84/Biosolid.pumpkin.row.jpg *License:* CC BY 3.0 *Contributors:* Own work *Original artist:* Red58bill

- **File:Blue-green_algae_cultured_in_specific_media.jpg** *Source:* https://upload.wikimedia.org/wikipedia/commons/1/18/Blue-green_algae_cultured_in_specific_media.jpg *License:* CC BY-SA 3.0 *Contributors:* Own work *Original artist:* Joydeep

- **File:Bokashi_bin_-_inside.JPG** *Source:* https://upload.wikimedia.org/wikipedia/commons/5/5b/Bokashi_bin_-_inside.JPG *License:* CC BY-SA 3.0 *Contributors:* Own work *Original artist:* Pfctdayelise

- **File:Bomberos2.jpg** *Source:* https://upload.wikimedia.org/wikipedia/commons/f/f9/Bomberos2.jpg *License:* CC BY-SA 3.0 *Contributors:* Own work *Original artist:* Ramiro Barreiro

- **File:CSIRO_ScienceImage_4522_Testing_for_pathogens_in_agricultural_soil_containing_biosolids.jpg** *Source:* https://upload.wikimedia. org/wikipedia/commons/c/cd/CSIRO_ScienceImage_4522_Testing_for_pathogens_in_agricultural_soil_containing_biosolids.jpg *License:* CC BY 3.0 *Contributors:* http://www.scienceimage.csiro.au/image/4522 *Original artist:* David McClenaghan, CSIRO

- **File:Chicken_Sheds_at_Balado_Airfield_-_geograph.org.uk_-_354043.jpg** *Source:* https://upload.wikimedia.org/wikipedia/commons/ f/f1/Chicken_Sheds_at_Balado_Airfield_-_geograph.org.uk_-_354043.jpg *License:* CC BY-SA 2.0 *Contributors:* From geograph.org.uk *Original artist:* Brendan Hamill

- **File:Coal_miner_spraying_rock_dust.jpg** *Source:* https://upload.wikimedia.org/wikipedia/commons/e/ea/Coal_miner_spraying_rock_dust. jpg *License:* Public domain *Contributors:* http://www.flickr.com/photos/niosh/8743393753/ *Original artist:* National Institute for Occupational Safety and Health

- **File:Commons-logo.svg** *Source:* https://upload.wikimedia.org/wikipedia/en/4/4a/Commons-logo.svg *License:* ? *Contributors:* ? *Original artist:* ?

- **File:Compost_pile.JPG** *Source:* https://upload.wikimedia.org/wikipedia/commons/8/86/Compost_pile.JPG *License:* CC0 *Contributors:* Own work *Original artist:* Ksd5

- **File:Compost_site_germany.JPG** *Source:* https://upload.wikimedia.org/wikipedia/commons/e/e5/Compost_site_germany.JPG *License:* CC-BY-SA-3.0 *Contributors:* Own work *Original artist:* Crystalclear

- **File:Composting_in_the_Escuela_Barreales.jpg** *Source:* https://upload.wikimedia.org/wikipedia/commons/7/74/Composting_in_the_Escuela_ Barreales.jpg *License:* Public domain *Contributors:* Transferred from en.wikipedia; transferred to Commons by User:Liftarn using CommonsHelper. *Original artist:* Original uploader was Diego Grez at en.wikipedia

- **File:Corn_01.JPG** *Source:* https://upload.wikimedia.org/wikipedia/commons/0/03/Corn_01.JPG *License:* CC BY-SA 3.0 *Contributors:* Transferred from ml.wikipedia by Sreejith K (<a href='//commons.wikimedia.org/wiki/User_talk:Sreejithk2000' title='User talk:Sreejithk2000'>talk</a>) *Original artist:* Original uploaded by Ashlyak.

- **File:End_point_(4315712587).jpg** *Source:* https://upload.wikimedia.org/wikipedia/commons/f/f4/End_point_%284315712587%29.jpg *License:* CC BY 2.0 *Contributors:* End point *Original artist:* Clarity J from Canada

- **File:Feannaganairighaird.jpg** *Source:* https://upload.wikimedia.org/wikipedia/commons/b/bd/Feannaganairighaird.jpg *License:* CC BY-SA 2.0 *Contributors:* This image was taken from the Geograph project collection. See this photograph's page on the Geograph website for the photographer's contact details. The copyright on this image is owned by **Bob Embleton** and is licensed for reuse under the Creative Commons Attribution-ShareAlike 2.0 license. *Original artist:* See above.

- **File:Fish_Meal_Factory,_Westfield,_West_Lothian.jpg** *Source:* https://upload.wikimedia.org/wikipedia/commons/7/70/Fish_Meal_Factory% 2C_Westfield%2C_West_Lothian.jpg *License:* CC BY-SA 2.0 *Contributors:* Geograph http://www.geograph.org.uk/photo/100024 *Original artist:* paul birrell

- **File:Fish_meal_factory_at_the_north-west_corner_of_Bressay.jpg** *Source:* https://upload.wikimedia.org/wikipedia/commons/2/2b/Fish_ meal_factory_at_the_north-west_corner_of_Bressay.jpg *License:* CC BY-SA 2.0 *Contributors:* From www.geograph.org.uk *Original artist:* Colin Smith

- **File:Folder_Hexagonal_Icon.svg** *Source:* https://upload.wikimedia.org/wikipedia/en/4/48/Folder_Hexagonal_Icon.svg *License:* Cc-by-sa-3.0 *Contributors:* ? *Original artist:* ?

- **File:Food-scraps-compost.jpg** *Source:* https://upload.wikimedia.org/wikipedia/commons/b/b2/Food-scraps-compost.jpg *License:* CC BY 2.0 *Contributors:* Flickr: compost *Original artist:* Philip Cohen

- **File:France_Loiret_La_Bussiere_Potager_05.jpg** *Source:* https://upload.wikimedia.org/wikipedia/commons/f/f0/France_Loiret_La_Bussiere_ Potager_05.jpg *License:* CC-BY-SA-3.0 *Contributors:* No machine-readable source provided. Own work assumed (based on copyright claims). *Original artist:* No machine-readable author provided. Calips assumed (based on copyright claims).

- **File:Green_compost_bin.JPG** *Source:* https://upload.wikimedia.org/wikipedia/commons/d/db/Green_compost_bin.JPG *License:* CC0 *Contributors:* Own work *Original artist:* Siddharth Patil

- **File:Heogan_Fishmeal_Factory,_Bressay_-_geograph.org.uk_-_1444467.jpg** *Source:* https://upload.wikimedia.org/wikipedia/commons/ c/ce/Heogan_Fishmeal_Factory%2C_Bressay_-_geograph.org.uk_-_1444467.jpg *License:* CC BY-SA 2.0 *Contributors:* From geograph.org.uk *Original artist:* Mike Pennington

- **File:Hestemøj.jpg** *Source:* https://upload.wikimedia.org/wikipedia/commons/c/c2/Hestem%C3%B8j.jpg *License:* CC-BY-SA-3.0 *Contributors:* Own work *Original artist:* Malene Thyssen

- **File:Hjul-_eller_kärrårder,_Nordisk_familjebok.png** *Source:* https://upload.wikimedia.org/wikipedia/commons/5/57/Hjul-_eller_k%C3% A4rr%C3%A5rder%2C_Nordisk_familjebok.png *License:* Public domain *Contributors:* Nordisk familjebok (1922), vol.33, p.1107-1108 [1] *Original artist:* Nordisk familjebok

- **File:HomeComposting_Roubaix_Fr59.JPG** *Source:* https://upload.wikimedia.org/wikipedia/commons/b/b3/HomeComposting_Roubaix_ Fr59.JPG *License:* CC BY-SA 3.0 *Contributors:* Own work *Original artist:* F Moreau Lille3

- **File:INTERIOR,_DETAIL_OF_MANURE_CAR_-_Ritter_Ranch,_Chicken_House,_Dolores,_Montezuma_County,_CO_HABS_ COLO,42-DOL.V,4D-2.tif** *Source:* https://upload.wikimedia.org/wikipedia/commons/a/ac/INTERIOR%2C_DETAIL_OF_MANURE_CAR_ -_Ritter_Ranch%2C_Chicken_House%2C_Dolores%2C_Montezuma_County%2C_CO_HABS_COLO%2C42-DOL.V%2C4D-2.tif *License:* Public domain *Contributors:* http://www.loc.gov/pictures/item/co0136.photos.021735p *Original artist:* ?

25.5.3   Content license

www.ingramcontent.com/pod-product-compliance
Lightning Source LLC
Chambersburg PA
CBHW080829180526
45168CB00006B/2618